NEUROSCIENCE RESEARCH PROGRESS

CALCIUM SIGNALING AND NERVOUS SYSTEM

OVERVIEW AND DIRECTIONS FOR RESEARCH

NEUROSCIENCE RESEARCH PROGRESS

Additional books and e-books in this series can be found on Nova's website under the Series tab.

NEUROSCIENCE RESEARCH PROGRESS

CALCIUM SIGNALING AND NERVOUS SYSTEM

OVERVIEW AND DIRECTIONS FOR RESEARCH

VICTOR V. CHABAN

Copyright © 2020 by Nova Science Publishers, Inc.

All rights reserved. No part of this book may be reproduced, stored in a retrieval system or transmitted in any form or by any means: electronic, electrostatic, magnetic, tape, mechanical photocopying, recording or otherwise without the written permission of the Publisher.

We have partnered with Copyright Clearance Center to make it easy for you to obtain permissions to reuse content from this publication. Simply navigate to this publication's page on Nova's website and locate the "Get Permission" button below the title description. This button is linked directly to the title's permission page on copyright.com. Alternatively, you can visit copyright.com and search by title, ISBN, or ISSN.

For further questions about using the service on copyright.com, please contact:
Copyright Clearance Center
Phone: +1-(978) 750-8400   Fax: +1-(978) 750-4470   E-mail: info@copyright.com.

NOTICE TO THE READER

The Publisher has taken reasonable care in the preparation of this book, but makes no expressed or implied warranty of any kind and assumes no responsibility for any errors or omissions. No liability is assumed for incidental or consequential damages in connection with or arising out of information contained in this book. The Publisher shall not be liable for any special, consequential, or exemplary damages resulting, in whole or in part, from the readers' use of, or reliance upon, this material. Any parts of this book based on government reports are so indicated and copyright is claimed for those parts to the extent applicable to compilations of such works.

Independent verification should be sought for any data, advice or recommendations contained in this book. In addition, no responsibility is assumed by the Publisher for any injury and/or damage to persons or property arising from any methods, products, instructions, ideas or otherwise contained in this publication.

This publication is designed to provide accurate and authoritative information with regard to the subject matter covered herein. It is sold with the clear understanding that the Publisher is not engaged in rendering legal or any other professional services. If legal or any other expert assistance is required, the services of a competent person should be sought. FROM A DECLARATION OF PARTICIPANTS JOINTLY ADOPTED BY A COMMITTEE OF THE AMERICAN BAR ASSOCIATION AND A COMMITTEE OF PUBLISHERS.

Additional color graphics may be available in the e-book version of this book.

Library of Congress Cataloging-in-Publication Data

ISBN: 978-1-53618-403-7

Published by Nova Science Publishers, Inc. † New York

CONTENTS

Preface vii

Introduction xi

Chapter 1 Role of Calcium Signaling in Peripheral Sensitization of Primary Afferent Sensory Neurons 1

Chapter 2 Role of Group II Metabotropic Glutamate Receptors in Estrogen Attenuations of ATP-Induced Ca2+ Signaling in Sensory Neurons 7

Chapter 3 Interaction between Purinergic (P2X3) and Estrogen (ERa /ERβ) Receptors in ATP-Mediated Calcium Signaling in Sensory Neurons 13

Chapter 4 Calcium Communication between Neuroblastoma SH-SY5Y and Primary Sensory Neurons 19

Chapter 5 Calcium Signaling and Sensory Neuronal Reorganization 31

Chapter 6 Role of TRPV1 and P2X2/3 Receptors in Ca2+– Mediated Nociception in Visceral Sensory Neurons 37

Chapter 7 Calcium Signaling in Astrocytes 53

Chapter 8	Calcium Signaling in Visceral Nociception	63
Chapter 9	Calcium Signaling in Chronic and Neuropathic Pain	71
Chapter 10	Calcium Signaling and Regulation of Brain Function	79
About the Author		89
Index		91

PREFACE

Calcium signaling is a general process of communication for all cells that balances variety of receptors, pumps, transporters and channels for subsequent intracellular transduction in normal and pathological states. It includes plasma membrane bound channels as well as intracellular receptors distributed in distinct cellular and tissue compartments that act to maintain physiological effects. Ca^{2+} is the fifth most abundant element in the body that plays vital role in many physiological and pathophysiological states. Ca^{2+} balance is critical in normal lifespan and aging with children are in positive Ca^{2+} homeostatic balance, healthy adults in neutral and elderly individuals are typically in negative. Integrated hormonal system mainly controls Ca^{2+} homeostasis within the gut, musculoskeletal system and kidney. Several Ca^{2+} channels such as voltage-gated, store- operated and ligand- gated have been identified with clear differences in their influence of particular cellular response. This extremely complex channel system contributes to almost all aspects of homeostasis by regulating neurotransmission, muscle contraction, metabolism, growth, motility, fertilization and cell death.

The idea that calcium plays significant role in the maintenance of cellular structure is going back more than century ago when Sydney Ringer established the importance, both qualitative and quantitative, of calcium ions in the contraction of the heart [1]. The way of present understanding of calcium homeostasis has been long and complicated. Pioneering work of 1963 Nobel Prize winners Alan Hodgkin and Andrew Huxley signified key role of Ca^{2+} channels in cellular signaling. The Hodgkin-Huxley model established a framework and functional properties of these on channels, their selectivity, and gating [2, 3]. Another classic manuscript by Bernard Katz and Ricardo Miledi proposed the "calcium hypothesis," which states that Ca^{2+} entry into the axon terminal is an essential component of the depolarization-release coupling process [4]. For many years, scientific team lead by late Platon Kostyuk where author completed his doctoral studies contributed significantly to our present understanding of the calcium signaling in excitable cells showing that this is one of the main coupling mechanism linking external stimuli, via the membrane, with intracellular cascades [5]. From author's early work it is known that calcium-selective voltage- operated channels form a main pathway for intracellular signal transduction, however the diversity of Ca^{2+} channels is well evidenced.

A natural way to differentiate several types of calcium channels would be to modulate their activity by different known transmitters, hormones or modulators (such as agonists and antagonists). Today, calcium is one of the most studied second messenger ion within the cell but its function is greatly depends on cell or tissue type. A defect or absence of calcium signaling component induces homeostatic compensatory cascade by activating calcium homeostasome leading to the reorganization of the entire system. A better understanding of calcium homeostasome may help to unravel the details and causes for many metabolic and functional syndromes and in turn lead to the development of better therapeutic interventions to combat the clinical symptoms associated with particular disease. This book will summarize past and recent author's work in the area of calcium signaling in neuronal and glial cells such as Ca^{2+} influx into the cell, buffering in the cytosol, as well as newly identified pharmaceutical tools for Ca^{2+} modulation and the involvement of calcium signals in health and progression of disease.

Dr. Victor V. Chaban, Professor of Medicine
Charles R. Drew University and University of California Los Angeles

REFERENCES

[1] Ringer S. Concerning the influence exerted by each of the constituents of the blood on the contraction of the ventricle, *Journal of Physiology*, 1882 Aug; 3(5-6): 380–393.

[2] Hodgkin A. The Croonian Lecture: Ionic movements and electrical activity in giant nerve fibres. *Proceeding Royal Society* (London), 1958, 148: 1–37.

[3] Hodgkin A. and Huxley A. A quantitative description of membrane current and its application to conduction and excitation in nerve. *Journal of Physiology*.1952; 117: 500– 544.

[4] Katz B., Miledi R. The timing of calcium action during neuromuscular transmission. *Journal of Physiology*, 1967, 189: 535-544.

[5] Kostyuk P. and Verkhratsky A. Calcium Signaling in the Nervous System, 1995, Wiley, 206 p.

INTRODUCTION

This book represents author's work for over two decades on calcium signaling in signal transduction of both excitable and non-excitable cells. The research studies in which I have been involved at University of California Los Angeles (UCLA) and the Charles R. Drew University of Medicine and Science (CDU) are part of the large-scale effort aimed to discover treatment of many chronic diseases characterized by changes in calcium signaling. An important step in the clarification of the mechanisms of calcium signaling was made by the development of intracellular calcium imaging technique as well as voltage- and patch- clamp. Next logistical step was to answer the question; what happened to Ca^{2+} ions inside the cell during its interaction with various second messenger systems? With my collaborators, we characterized temporal and spatial organization of Ca^{2+} pathways in primary afferent neurons located in dorsal root ganglia (DRG) innervating the intestine, reproductive viscera and somatic tissues.

The long-stated (55 years) "Gate Control Theory of Pain" seeks to explain why non-painful sensory input limits pain information from reaching the central nervous system. The focus of this theory is at the level of the spinal cord [1]. In this book I was seeking to understand a novel form of pain filtering that resides within dorsal root ganglia (DRG). The canonical view of DRG function is one of biochemical and molecular support; it is where the cell bodies and nuclei of sensory neurons reside.

There is a strong body of experimental and theoretical work that supports an integrative role, and the hypotheses proposed here reflect new considerations and approaches to the integration of action potential conduction [2]. Studies, described here intended to develop a coherent framework for our understanding of role of Ca^{2+} signaling within the DRG, heterogeneity of nociceptive neurons as a as a gate for pain information and the development of novel therapeutic tools for pain relief. Specifically, our studies helped to define the role calcium changes in ATP-sensitive purinergic (P2X) and NMDA receptor activation and the role of group II metabotropic glutamate receptors channels in estrogen- modified cellular pathways. The novel concept of role of glial cells in calcium signaling changed significantly from previous view of these cells as a passive elements in nervous system predominantly through the activation of potassium channels to the ability to generate complex and organized Ca^{2+} signals.

Pain is a complex and personal experience. Pain strongly implicated in the etiology of many diseases, which often are complicated by co-morbid depression, panic and other psychiatric disorders, all pose health risks. Designing new and specific anti-nociceptive therapies will have a major impact on health-related quality of life in patients, significantly reducing therapeutic interventions. Together with my collaborators from the UCLA Department of Neurobiology and Obstetric and Gynecology as well as the Department of Anesthesiology, Harbor-UCLA Medical Center we performed cutting edge studies about the role of calcium changes in estrogen receptors modulation during nociception.

Calcium signaling contributes to the neuronal network capacity to change leading to neuroplasticity, key process in brain development and human existence [3]. Neuronal connectome forms the basis for communication and reorganization associated with homeostatic balance or cognitive development. Surely, Ca^{2+} signaling is implicated in most aspects of cellular functioning, the mechanisms of which still not entirely understood: for example how calcium binding to specific intracellular proteins trigger the corresponding function or aging and death. From early embryogenesis to end stage of cell lifespan with "Ca^{2+} wave of life" (the explosive increase in calcium that occurs in the cytosol at fertilization) and "Ca^{2+} wave of death" (calcium-propagated necrotic wave that promotes organismal death) this signaling not only brings closer science, philosophy and religion but eventually lead to paradigm shift in our understanding of principles of life.

REFERENCES

[1] Melzack, R. and Wall P. Pain mechanisms: a new theory. *Science*, 1965. 150, 3699: p. 971- 979.
[2] Chaban V. *Unraveling the Enigma of Visceral Pain,* 2016, Nova Publishers, 90p.
[3] Chaban V. (Editor). *Neuroplasticity.* Insights of Neuronal Reorganization, InTech, 2018, 193p.

Chapter 1

ROLE OF CALCIUM SIGNALING IN PERIPHERAL SENSITIZATION OF PRIMARY AFFERENT SENSORY NEURONS

ABSTRACT

To understand the mechanisms of nociception is an important step in treating pain. During inflammation, increased nociceptive input from an inflamed organ can sensitize neurons that receive convergent input from an unaffected organ, but the site of visceral cross-sensitivity is unknown. We examined the cellular responses to ATP and substance P stimulation on intracellular calcium concentration ($[Ca^{2+}]_i$) in sensory neurons innervating visceral organs. Lumbosacral dorsal root ganglia (L6- S1) were cut into slices and processed for substance P receptor expression using immunocytochemistry. Primary culture of dorsal root ganglion (DRG) neurons was used for $[Ca^{2+}]_i$ measurement by videomicroscopy. Brief addition of low dose adenosine triphosphate (ATP, 5 mM) and substance P (10 mM) significantly increased $[Ca^{2+}]_i$ after subsequent ATP stimulation at the same neuron. Sensitization of the DRG neurons innervating the different organs may be through the release of nociceptive transmitters such as ATP and/or substance P within the ganglion as evidenced by their effect on intracellular calcium. Together, these experiments will increase our understanding of the important modulatory role of peripheral sensitization on calcium signaling during nociceptive transmission.

CALCIUM AND NOCICEPTION

The cell bodies of primary visceral spinal afferent neurons are located in the dorsal root ganglia (DRG). Primary afferents transmit information about chemical or mechanical stimulation from the periphery to the spinal cord. Nociceptors believe to be are small-to-medium size DRG neurons whose peripheral processes detecting potentially damaging physical and chemical stimuli. Adenosine triphosphate (ATP) has emerged as a putative signal for visceral pain. ATP is released by distention of the viscera and tissue damage. Nociceptive c-fibers that are activated by ATP and excitatory amino acids released by noxious stimuli from cells in target organs (paracrine action), from afferent terminals (autocrine action 2). We observed before that DRG neurons innervating viscera have a greater $[Ca^{2+}]_i$ response to subsequent N-methyl-D-aspartic acid (NMDA) stimulation than somatic afferents [1]. This observation indicates that these neurons express receptors with higher permeability to Ca^{2+}, which modulates transduction of nociceptive signals. Visceral DRG also express nociceptive ATP-sensitive purinergic (P2X3) and capsaicin-sensitive vanilloid (TRPV1) receptors. Sensitization of primary afferent neurons to stimulation may play a role in the enhanced perception of visceral sensation and pain. Chest pain from coronary heart disease, endometriosis, acute and recurrent/chronic pelvic pain in women or abdominal pain from bowel syndromes are all visceral pain sensations that may result in part from sensitization described above.

INTRACELLULAR CALCIUM $[CA^{2+}]_I$ FLUORESCENCE IMAGING

In our experiments for Ca2+ imaging analysis DRG neurons were loaded with fluorescence indicator dye Fura-2AM for 1 hour at 37° C in Hank's buffered saline solution (HBSS) containing 20 mM HEPES, pH 7.4 Coverslips were mounted in fast-perfusion chamber P-4 (World Precision

Instruments, Sarasota, Fl) and placed on a stage of Olympus IX51 inverted microscope. Observations were made at room temperature with 203 UApo/340 objective. A fast superfusion system was used to perfuse the cells with HBSS and rapidly apply ATP and substance P. Fluorescence intensity at 505 nm with excitation at 334 nm and 380 nm was captured as digital images (sampling rates of 0.1–2 s). Regions of interest were identified within the soma from which quantitative measurements were made by re-analysis of stored image sequences using Slidebook Digital Microscopy software. $[Ca^{2+}]_i$ was determined by ratiometric method of Fura-2 fluorescence from calibration of series of buffered Ca^{2+} standards. ATP and substance P were acutely applied onto the experimental chamber. Repeated application of drugs was achieved by superfusion in a rapid mixing chamber into individual neurons for specific intervals (100–500 ms). Cells were perfused with experimental media (2 ml/ min) using a Rainin peristaltic pump.

ATP-INDUCED CA^{2+} FLUX IN DRG NEURONS

Lumbosacral DRG were collected from rats and cultured for 48–72 hrs on coverslips before being loaded with Fura-2 acetoxymethyl ester (Fura2AM, Molecular Probes, Eugene, OR) by incubating cultures in a 5 mM solution for 1 hour at 37°C. Stimulation of small to medium size DRG neurons (35 mm in diameter) with 10 mM ATP caused transient $[Ca^{2+}]_i$ increase (207.6 621.8 nM) in about 50% of neurons tested (n=510). Repeated ATP (10 mM) application by fast superfusion (10 sec) produced reproducible $[Ca^{2+}]_i$ responses (Figure 1.1). The response to ATP was fully reversed after a 10–15 min washout with Hank's Balanced Salt Solution (HBSS; Gibco BRL buffered with 20 mM N- (2- hydroxyethyl) piperazine-N-(2-ethanesulfonic) acid: HEPES). Chelating extracellular Ca^{2+} with BAPTA (1,2- bis(o-aminophenoxy)ethane-N,N,N9, N9-tetraacetic acid, 10 mM) eliminated ATP-induced $[Ca^{2+}]_i$ responses. Pretreatment with the purinoreceptor antagonist phosphate-6-azophenyl-29,49- disulfonic acid (PPADS) inhibited ATP responses: 5 mM produced a 70% inhibition and

10 mM completely blocked the ATP-induced $[Ca^{2+}]_i$. These results indicate that ATP-induced Ca^{2+} flux is dependent on activation of P2X receptors and plasma membrane Ca^{2+} channels, which sensitize the DRG cells to repeated ATP stimulation.

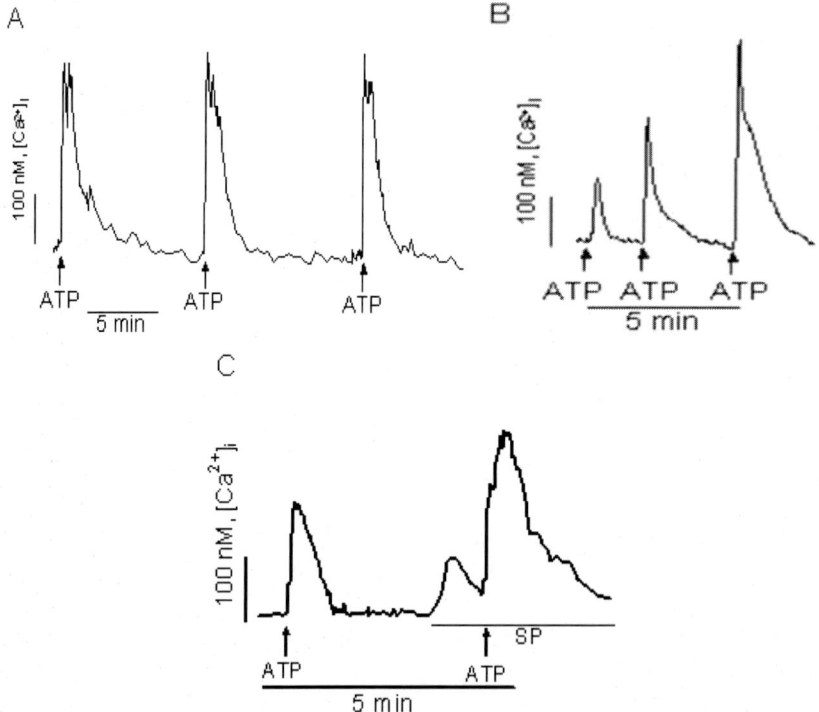

Figure 1.1. Sensitization of DRG neurons in vitro. A) Repeated ATP (10 μM) stimulation after 10 min wash-out with experimental medium produced reproducible $[Ca^{2+}]_i$ responses in DRG neurons (time and $[Ca^{2+}]_i$ indicated by bars). The trace is from a single neuron and shows features of typical responses with rapid rise after stimulation (indicated by arrow) followed by a slower recovery; B) Sensitization of $[Ca^{2+}]_i$ response to repeated stimulation with ATP (5 μM); C) Substance P (SP; 10 μM) increases ATP-induced $[Ca^{2+}]_i$ flux suggesting sensitization. Adopted from [2].

Our data suggest that ATP plays a vital role in cellular metabolism through Ca^{2+} signaling. ATP transduces noxious stimuli by activating purinergic, ATP-gated P2X receptors on primary afferent fibers. Opening of P2X channels results in membrane depolarization sufficient to trigger action potentials and Ca^{2+} influx through defined voltage-gated calcium channels (VGCC) associated with nociception. According to this theory, pain of tissue irritation (mechanical distortion or inflammation) is due to ATP activation of high threshold nociceptors [3]. Significantly, inflammation dramatically alters purinoception by causing a several fold increase in ATPactivated currents, alters the voltage dependence of P2X receptors, and enhances the expression of P2X receptors increasing neuronal hypersensitivity.

The inflammatory process produces a number of mediators, which activate nociceptors by interacting with ligand-gated ion channels or by sensitizing primary afferents. One mechanism for sensitization involves phosphorylation of ion channels and receptors including P2X and substance P. Inflammation does not change the percentage of total cells responding to ATP but sensitizes the ATP response and increases the expression of P2X2 and P2X3 [4]. Thus, the increased sensitivity during the inflammation is due to an increase in ATP responses suggesting that a small amount of ATP would evoke depolarization sufficient to elicit action potentials in DRG neurons. This pathological response arises from sensitization of DRG neurons to external stimuli. Previous studies have demonstrated that a large percentage of visceral afferent fibers are substance P-positive (50%) [5]. Capsaicin treatment abolished virtually all the immunoreactive fibers suggesting that these afferents also express nociceptive TRPV1 receptors.

DRG neurons in vitro are a well-accepted model to examine primary afferent response to nociceptive and anti-nociceptive signals. It should also be noticed that nociceptive systems implicated in the etiology of many pain-associated disorders might be complicated by input of other signaling pathways such as endogenous opioids and others. Within the context of our hypothesis, modulation of nociceptive response depends on the type of pain, its durations and the involvement of other anti-nociceptive mechanisms. In this study we investigated the mechanisms of sensitization of DRG neurons in response to application of substance P in ATP-evoked Ca^{2+}- transients. Our data suggest that either sub-threshold dose of ATP or pretreatment with substance P produce enhanced Ca^{2+} response leading to the sensitization. Sensitization accounts for a dramatically lowered nociceptive threshold to mechanical manipulation of the inflamed area. Within the context of the cross-sensitization hypothesis, inflammation sensitizes non-inflamed viscera that are innervated by the same DRG. Cross-sensitization occurs as a result of intra-DRG release of sensitizing mediators such as ATP or substance P, which we modeled in cultured DRG by repeated ATP or substance P and ATP stimulation. Visceral nociception and nociceptor sensitization appear to be regulated by ATP and substance P, thus, the DRG is an important site of visceral afferent convergence and cross-sensitization.

REFERENCES

[1] McRoberts JA, Coutinho SV, Marvizon JC, et al.. Role of peripheral N-methyl-D-aspartate (NMDA) receptors in visceral nociception in rats. *Gastroenterology.* 2001; 120 (7): 1737– 1748.
[2] Chaban V. (2016) *Unraveling the Enigma of Visceral Pain,* Nova Publishers, New York, 77p.
[3] Bodin P, Burnstock G. Purinergic signalling: ATP release. *Neurochemical Research.* 2001; 26(8–9): 959–969.
[4] Xu GY, Huang LY. Peripheral inflammation sensitizes P2X receptor-mediated responses in rat dorsal root ganglion neurons. *Journal of Neurosciences.* 2002; 22(1): 93–102.
[5] Papka RE, Storey-Workley M, Shughrue PJ, et al. Estrogen receptor-alpha and beta- immunoreactivity and mRNA in neurons of sensory and autonomic ganglia and spinal cord. *Cell Tissue Research.* 2001; 304(2): 193–214.

Chapter 2

ROLE OF GROUP II METABOTROPIC GLUTAMATE RECEPTORS IN ESTROGEN ATTENUATIONS OF ATP-INDUCED CA^{2+} SIGNALING IN SENSORY NEURONS

ABSTRACT

Estrogen attenuation of ATP-induced increase of intracellular calcium concentration ($[Ca^{2+}]_i$) in rat dorsal root ganglion (DRG) neurons was studied to explore an estradiol regulation of pain. WE showed that 17β-estradiol (E2) has been shown to modulate L-type VGCC through a membrane estrogen receptor-group II metabotropic glutamate receptor (mGluR2/3). Previously we showed that DRG (L1-S3) express mGluR2/3 receptors. DRG neurons were stimulated twice, once with ATP (50 μM) for 5 seconds, and then again in the presence of E2 (100 nM) or E2 (100 nM) + LY341495 (100nM), an mGluR2/3 inhibitor to check changes in $[Ca^{2+}]_i$. ATP produced a transient increase in $[Ca^{2+}]_i$ (216.3 ± 41.2 nM). This transient could be evoked several times in the same DRG neurons if separated by a 5 min washout. Treatment with E2 significantly attenuated the ATP-induced $[Ca^{2+}]_i$ in 60% of the DRG neurons, to 163.3±20.9 nM ($p<0.001$). Co-application of E2 and the mGluR2/3 inhibitor LY341495 blocked the 17β-estradiol-attenuation of the ATP-induced $[Ca^{2+}]_i$ transient (209.1±32.2 nM, $p>0.05$). These data indicated that the rapid action of E2 in DRG neurons is dependent on the mGluR2/3, and demonstrate that membrane estrogen receptor initiated signaling involves tan interaction with mGluRs.

ROLE OF G-PROTEIN-COUPLED SIGNALING IN INTRACELLULAR CALCIUM CHANGES

Adult DRG neurons in short-term culture continue to respond to estrogen receptors (ERs) agonists mimicking *in vivo* activation. Many of the E2-mediated signaling processes have been ascribed as membrane-associated. E2 modulates cellular activity by altering ion channel opening, G-protein signaling, and activation of trophic factor-like signal transduction pathways [1]. However, the mechanism(s) by which ERs are able to trigger cell signaling when localized at the neuronal membrane surface not yet fully determined. Previous data indicate that membrane-associated ER directly interacts with mGluRs to mediate intracellular signaling [2]. DRG neurons express mGluR2/3 receptors indicating that glutamate could have a substantial inhibitory effect of primary afferent function, reducing and/or fine-tuning sensory input before transmission to the spinal cord. Here we addressed the question of whether ERs interact with group II metabotropic glutamate receptors by using well-established cellular model to study the effect of blockade of mGluR2/3 on E2-mediated ATP-induced Ca^{2+} signaling in small DRG neurons to help achieve a deeper understanding of this mechanism.

In our experiments we investigated whether the rapid E2 attenuation of ATP-induced increase $[Ca^{2+}]_i$ requires the mGluR2/3. DRGs were stimulated twice, once with ATP (50 μM) for 5 seconds, and then again in the presence of E2 (100 nM) or E2 (100 nM) + LY341495 (100nM), a mGluR2/3 inhibitor. Treatment with E2 significantly attenuated the ATP-induced $[Ca^{2+}]_i$ in 65% of the DRG neurons, to 163.3±20.9 nM vs 216.3 ±41.2 nM in control (p<0.001) (Figure 2.1A). Co-application of E2 and LY341495 blocked the E2 attenuation of the ATP-induced $[Ca^{2+}]_i$ transient (209.1±32.2 nM, p>0.05) (Figure 2.1 B). Together, these data indicate that the rapid action of E2 in DRG neurons is dependent on the mGluR2/3, and demonstrate that membrane estrogen receptor initiated signaling involves an interaction with mGluRs.

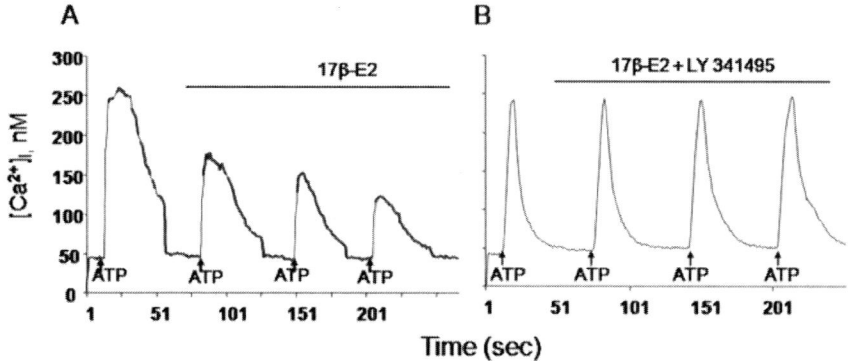

Figure 2.1. A. Treatment with 17β-estradiol (17β-E2) significantly attenuated the ATP-induced [Ca2+]i of the DRG neurons, from 216.3 ±41.2 nM to 163.3±20.9 nM ($p<0.05$) in 65% small DRG neurons. B. Co-application of 17β-estradiol and the mGluR2/3 inhibitor LY341495 blocked the estradiol-attenuation of the ATP-induced [Ca2+]i (209.1±32.2 nM) compared to control (216.3 ±41.2 nM; $p>0.05$).

Sex steroids have been suggested as a plausible mechanism of pain regulation and our results provide the evidence that estrogens may directly influence nociception at the level of primary sensory neurons by interacting with group II metabotropic glutamate receptors and suggest that lack of estrogens enhances pro-nociception. Several recent studies have shown that the peripheral nervous system is under substantial influence of gonadal hormones [3]. In this study we hypothesized that DRG neurons responded to E2 attenuation of ATP-induced $[Ca^{2+}]_i$ spikes through interaction with mGluR2/3 (Figure 2.2).

These results do not preclude an additional mechanism through which estradiol may be acting, such as in the presence of opioids, estrogen acts differently than by itself on nociceptive-mediated pathways. Some previous and current studies have established connections between estrogens and the modulation of different nociceptive pathways and that an estrogen can inhibit nociceptive stimulus such as ATP through interaction with mGluR2/3 [4]. Painful stimuli initiate action potentials in the peripheral terminals of DRG neurons evoking release of excitatory neurotransmitters such as glutamate.

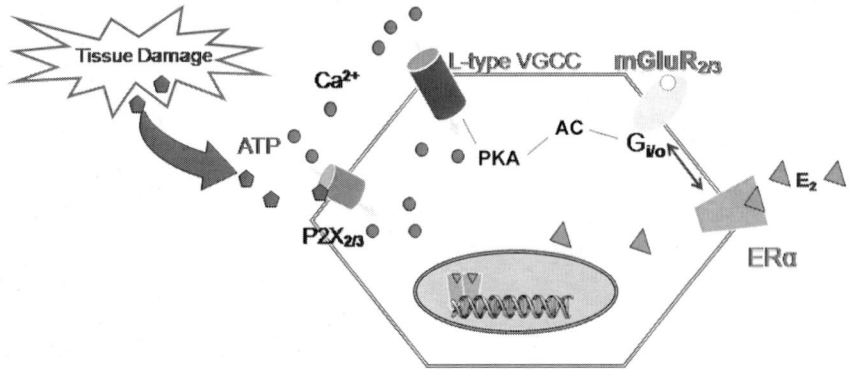

Figure 2.2. Proposed mechanism of estradiol effect on ATP-induced [Ca2+]i signaling in primary sensory neurons. ATP released by tissue damage acts on P2X3 that activate voltage-gated Ca2+ channels (VGCC)-mediated nociception signaling. 17β-E2 modulates L-type VGCC in DRG neurons via the direct interaction of a membrane ERα with the mGluR2/3 inhibiting adenylyl cyclase activation of L-type VGCC (Adapted from [5].

REFERENCES

[1] Kelly MJ, Levin ER. Rapid actions of plasma membrane estrogen receptors. *Trends in Endocrinology and Metabolism.* 2001; 12: 152–156.

[2] Dewing P, Boulware MI, Sinchak K, Christensen A, Mermelstein PG, Micevych P. Membrane estrogen receptor-alpha interactions with metabotropic glutamate receptor 1a modulate female sexual receptivity in rats. *Journal of Neuroscience.* 2007; 27: 9294–9300.

[3] Xu GY, Shenoy M, Winston JH, Mittal S, Pasricha PJ. P2X receptor-mediated visceral hyperalgesia in a rat model of chronic visceral hypersensitivity. *Gut.* 2008; 57:1230–1237.

[4] Tashiro A, Nishida Y, Bereiter D. Local group I mGluR antagonists reduce TMJ-evoked activity of trigeminal subnucleus caudalis neurons in female rats. *Neuroscience* 2015. 23; 299: 125-3.

[5] Chaban V, Li J, McDonald J, Rapkin A. and Micevych P. Estradiol attenuates ATP-induced increase of intracellular calcium through group II metabotropic glutamate receptors in rat DRG neurons. *Journal of Neuroscience Research.* 2011, 89 (11): 1707–1710.

Chapter 3

INTERACTION BETWEEN PURINERGIC (P2X3) AND ESTROGEN (ERA /ERB) RECEPTORS IN ATP-MEDIATED CALCIUM SIGNALING IN SENSORY NEURONS

ABSTRACT

Emerging evidence support a role of purinergic P2X3 receptors in modulating nociceptive signaling in sensory neurons. Lumbosacral DRG neurons (L1-S1) express both estrogen (ERα and ERβ) receptors. In this study we investigated the expression of P2X3 receptors and the effect of 17β-estradiol (E2) on ATP-induced $[Ca^{2+}]_i$ ncrease in DRG neurons collected from wild type C57Bl/6J, ERαKO and ERβKO knock-out mice. Our data showed a significant decrease for P2X3 in ERαKO (all levels) and ERβKO (mostly observed in L1, L2, L4, and L6). Furthermore, E2 (100 nM) significantly attenuated the ATP (10 μM)-induced $[Ca^{2+}]_i$ in C57Bl/6J mice. ERs antagonist ICI 182,780 (1μM) blocked this attenuation. Homomeric P2X3 receptors are plentifully expressed in DRG neurons and contribute to nociceptive signals. α,β-me ATP which is a specific agonist of P2X2/3 receptors showed similar responses to the ATP-induced calcium increase in knock-out mice. A membrane-impermeable E-6-BSA (1μM) had the same effect as E2 suggesting action on the membrane. In DRG neurons from ERβKO and WT miceE2 attenuated the ATP/α,β-me ATP- induced $[Ca^{2+}]_i$ fluxes but in DRG neurons from ERαKO mice, this hormone had no effect suggesting that this attenuation depends on membrane-associated ERα receptors.

EFFECT OF E2 AND PHARMACOLOGICAL PROFILE OF E2-MEDIATED MODULATION ON ATP-INDUCED INTRACELLULAR CALCIUM CONCENTRATION ($[CA^{2+}]_I$) IN DRG NEURONS

Our data suggest that small diameter DRG neurons (presumably nociceptors) expressed Ca^{2+}- sensitive purinergic (P2X) receptors (Figure 3.1). ATP-induced $[Ca^{2+}]_i$ transients in DRG neurons in mice, a result similar to that observed in rat DRG neurons [1]. Brief 10 second application of ATP (10μM) by fast superfusion produced equal $[Ca^{2+}]_I$ spikes in 65% of tested neurons.

After a 5-min washout with HBSS, additional stimulation with ATP (10μM) induced a subsequent $[Ca^{2+}]_i$ transients Pretreatment with purinergic receptor antagonist PPADS (5 μM) blocked the ATP-induced $[Ca^{2+}]_i$ transients. Similarly, ATP stimulation in a $[Ca^{2+}]_i$ free media with the Ca^{2+} chelator, BAPTA (10 mM), eliminated $[Ca^{2+}]_i$ spikes indicating the necessity for P2X3 receptors and extracellular Ca^{2+}. E2 (100 nM) by itself had no effect on basal $[Ca^{2+}]_i$ but this hormone attenuated the ATP-induced $[Ca^{2+}]_i$ transients. The effect of E2 was reversible. After the initial ATP response, five minute incubation with E2 inhibited ATP-induced $[Ca^{2+}]_i$ transient (440.3±58.3 vs. 280.4±48.8 nM, n=15, $p < 0.05$). To confirm the desensitization of E2, we administered application of repeated ATP, indicating that E2 does not desensitize upon of application of repeated ATP (Figure 3 (d). The estrogen receptor antagonist ICI 182,780 (1 μM) blocked the 17β-estradiol effect on attenuated ATP-induced $[Ca^{2+}]_i$ transients.

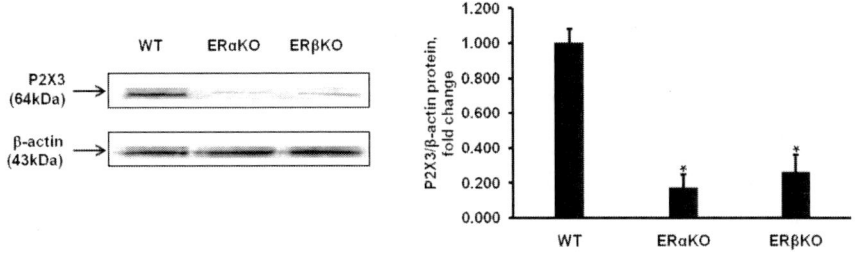

Figure 3.1. Western blot analysis of DRG lysates shows reduced expression of P2X3 in both knock-out mice. Quantification of signals from Western blots shows statistically significant difference between the intensity of the bands from both knock-out DRG neurons when compared with wild type animals (Adapted from [2]).

ESTROGEN (17B-ESTRADIOL) ATTENUATION OF ATP AND A,B-ME ATP-INDUCED INTRACELLULAR CALCIUM INFLUX ([CA^{2+}]$_I$) IN DRGS FROM KNOCK-OUT (ERαKO AND ERβKO) MICE

To confirm which ER subtype mediates the E2 attenuation of ATP-induced response, we compared estradiol action mediating $[Ca^{2+}]_i$ signaling in DRG neurons from Wt, and knockout mice. The effect of E2 was similar in ERβKO mouse DRG neurons to that observed in Wt mice (n = 125 cells/5 mouse) (Figure 3.2 b). However, in DRG neurons from ERαKO mice, E2 did not block ATP-induced $[Ca^{2+}]_i$ suggesting that its attenuation depends on ERα (n = 112 cells/5 mouse) (Figure 3.2 a). 17α-Estradiol had no effect on Wt, ERαKO or ERβKO mice. We also used α,β-me ATP, a specific agonist of P2X2/3, to confirm the observations presented with ATP (n=105 cells/each group 6 mouse). Even the properties of P2X2/3 receptors are not similar to those of the P2X3 receptors both P2X3 and P2X2/3 receptors may be a target of action for estradiol (Figure 3.3). Our experiments used a combination of techniques to determine that DRG neurons in culture can be used to study the cellular response to a putative nociceptive signal, ATP. Our data suggest that in primary DRG neuronal cultures, E2 attenuates the ATP/α,β-meATP -induced $[Ca^{2+}]_i$ responses and interferes with the membrane-associated ERα. In our experiments we noticed significant decrease in number of responsive neurons in both knock-out mice. Fewer than 20% of tested cells responded to ATP/α,β-meATP stimulation which correspond to the fact that that both ERKO and ERKO exhibit decrease in their expression of P2X3 receptors (Figure 3.1).

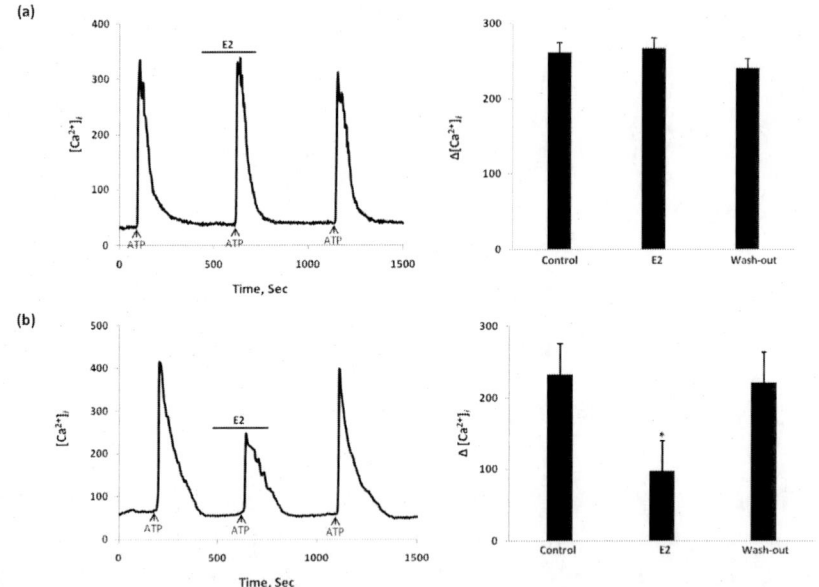

Figure 3.2. The effect of E2 on ATP-induced $[Ca^{2+}]_i$ transients in estrogen receptor-α knockout (ERαKO) and estrogen receptor-β knockout (ERβKO) mice. (a) In ERαKO mouse, E2 added for 5 min didn't inhibit ATP-induced $[Ca^{2+}]_i$ transient; (b) In ERβKO mouse E2 stimulation significantly attenuated the ATP-stimulated $[Ca^{2+}]_i$ transient similar to that observed in Wt mouse. Summary data represented on the right bar graphs. (Adopted from [2]).

The localization of ERs in DRG neurons [3] and the attenuation of ATP-induced $[Ca^{2+}]_i$ strongly suggest that E2 modulates of visceral nociception which appears to be regulated by ATP. More than half the small diameter DRG neurons (presumably nociceptors) in culture respond to ATP and were estrogen- sensitive. ATP effect was blocked by PPADS, indicating an involvement of purinoreceptors (Figure 3.1). The involvement of the P2X3 receptor in the ATP response was proven by using selective P2X3 agonist α,β-meATP (Figure 3.3) therefore ERα interacts with P2X3 in DRGs.

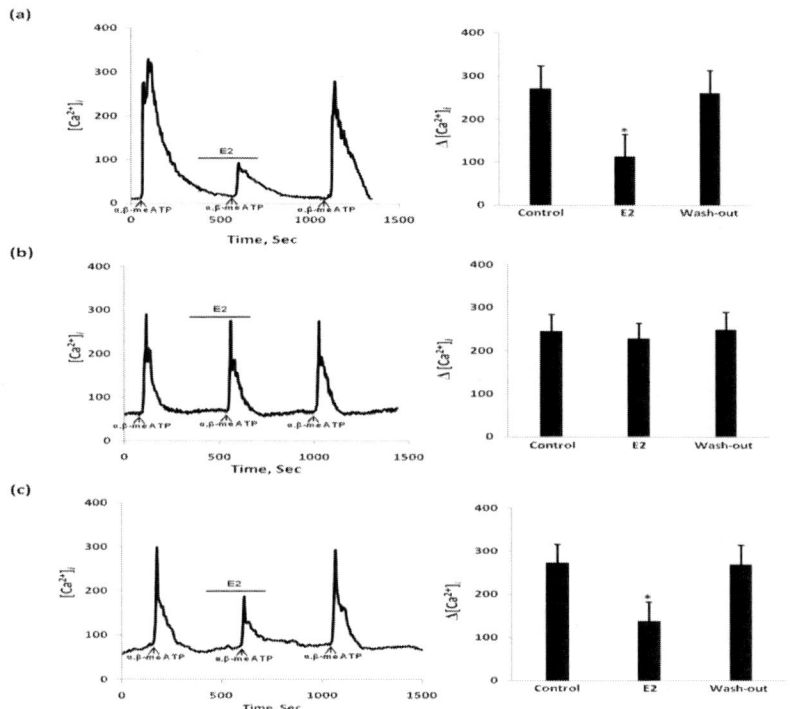

Figure 3.3. The effect of 17β-estradiol (E2) on α,β-meATP-induced $[Ca^{2+}]_i$ transients in Wt, ERαKO, and ERβKO mice. (a) Wt mouse as a control, (b) ERαKO mouse, E2 added for 5min didn't inhibit α,β-meATP-induced $[Ca^{2+}]_i$ transient in control vs. after E2 treatment. (c) In ERβKO mouse, E2 stimulation significantly attenuated the α,β-meATP-stimulated $[Ca^{2+}]_i$ transient similar to that observed in Wt mouse. Summary data represented on the right bar graphs. Values are expressed as mean±SEM. $\Delta[Ca^{2+}]_i$ were determined by subtracting the $[Ca^{2+}]_i$ peak levels from the basal $[Ca^{2+}]_i$ levels (Adapted from [2]).

REFERENCES

[1] Chaban V, Mayer E., Ennes H. and Micevych P. Estradiol inhibits ATP-induced intracellular calcium concentration increase in dorsal root ganglia neurons. *Neuroscience*. 2003; 118 (4): 941–948.

[2] Cho T. and Chaban V. Interaction between P2X3 and ERα/ERβ in ATP-mediated calcium signaling in mice sensory neurons, *Journal of Neuroendocrinology.* 2012, 24(5): 789–797.

[3] Papka R. and Mowa C. Estrogen receptors in the spinal cord, sensory ganglia, and pelvic autonomic ganglia. *International Review in Cytology.* 2003; 231: 91–127.

Chapter 4

CALCIUM COMMUNICATION BETWEEN NEUROBLASTOMA SH-SY5Y AND PRIMARY SENSORY NEURONS

ABSTRACT

Cell-cell communication occurs via a variety of mechanisms, including long distances (hormonal), short distances (paracrine and synaptic) or direct coupling via gap junctions, antigen presentation, or ligand-receptor interactions. We evaluated the possibility of neuro-hormonal independent, non-diffusible, physically disconnected pathways for cell-cell communication using dorsal root ganglion (DRG) neurons by measuring intracellular calcium $[Ca^{2+}]_i$ in primary culture DRG neurons. Physically disconnected (dish-in-dish system) mouse DRG were cultured for 12 hours near: a) media alone (control 1), b) mouse DRG (control 2), c) human neuroblastoma SHSY-5Y cells (cancer intervention), or d) mouse DRG treated with KCl (apoptosis intervention). We found that chemosensitive receptors $[Ca^{2+}]_i$ signaling did not differ between control 1 and 2. ATP (10 μM) and capsaicin (100nM) increased $[Ca^{2+}]_i$ transients to 425.86 + 49.5 nM, and 399.21 ± 44.5 nM, respectively. Estrogen (17β-estradiol (100 nM) exposure reduced ATP (171.17 ± 48.9 nM) and capsaicin (175.01±34.8 nM) $[Ca^{2+}]_i$ transients. The presence of cancer cells reduced ATP- and capsaicin-induced $[Ca^{2+}]_i$ by >50% (p<0.05) and abolished the 17β-estradiol effect. By contrast, apoptotic DRG cells increased initial ATP-induced $[Ca^{2+}]_i$ flux four fold and abolished subsequent $[Ca^{2+}]_i$ responses to ATP stimulation (p<0.001). Capsaicin (100nM) induced $[Ca^{2+}]_i$ responses were totally abolished. The local presence of apoptotic DRG or human neuroblastoma SHSY-5Y cells induced differing abnormal ATP and capsaicin-mediated $[Ca^{2+}]_i$ fluxes in normal DRG. These findings support physically disconnected, non-diffusible cell-to-cell calcium signaling. Further studies are needed to delineate the mechanism(s) of and model(s) of this novel communication.

CALCIUM SIGNALING IN THE PRESENCE OF PHYSICALLY ISOLATED NEURONAL CELLS

Despite remarkable advances in biomedical sciences and medical therapeutics in recent decades, the anticipated improvements in patient outcomes have not been realized owing to limited success in translating and widely diffusing scientific advances into clinical practice [1]. The promise of translational approaches can best be achieved by embracing the interdependence of basic science, clinical discovery, and patient-oriented research (clinical trials, bio-behavioral, community engagement, policy, etc.) and by testing new paradigms that might open previously unseen doors to accelerate improved patient outcomes [2]. One such door is the conventional conception of cellular communication, critical to clinical intervention out- comes, as driven mainly by direct coupling via gap junctions, antigen presentation, ligand- receptor interactions, or mediating diffusible factors. Likewise, cell-cell communication can occur over very short (paracrine and synaptic) or very long distances (hormonal). These modes of communication, however, all require either physical contact between cells or contact with mediating, diffusible factors. By contrast, while self - organization in biology has been viewed to be driven by diffusible signal mediated cell-cell communication, self- organization of dynamic collective systems events may occur in part, through a higher level of non-linear communication.

To study this phenomenon, we devised a "dish-in-dish" cell culture system to establish a framework for investigating whether physically disconnected cells could influence the behavior of other local cells in the absence of diffusible factors or other physical connection. We assessed invoked intracellular calcium currents ($[Ca^{2+}]_i$) in cultured DRG cells in response to ATP to activate purinergic receptors and capsaicin (vaniloid receptors agonist) in the presence or absence of estradiol, as the model of chemosensitive receptors $[Ca^{2+}]_i$ signaling. We posited that the local presence of dying cells and/or aberrantly behaving, e.g., cancer cells, were good candidates to express a form of communication that might alter the behavior of local, but normal cells. If physically disconnected neuro-hormonal independent, non-diffusible pathways do exist, then such influences should be detectable within cell signaling path- ways. Thus, we assessed $[Ca^{2+}]_i$ cell signaling responses in highly sensitive DRG neurons to well characterized chemosensitive receptors modulators in the presence of varying local, but physically isolated cells. Specifically, we investigated the $[Ca^{2+}]_i$ cell signaling responses in nor- mal physically isolated DRG neurons in primary culture in the local presence of media and no cells (control condition 1), the local presence of other normal DRG cells (control 2), and the local presence two distinct cell populations: human neuroblastoma SHSY-5Y cells (intervention 1) or DRG cells after KCl induced apoptosis (intervention 2) (Figure 4.1).

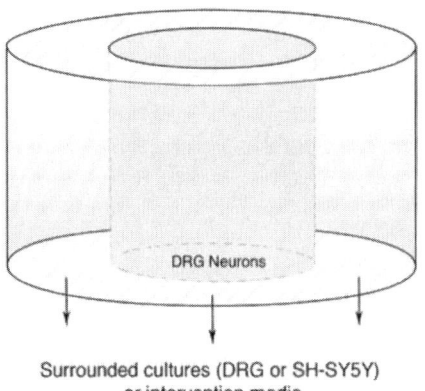

Figure 4.1. Experimental "dish in dish" system (U.S. Patent 10138451 (2018). DRG sensory neurons were plated and cultured in the "center" dish. Other DRG neurons, SH-SY5Y neuroblastoma cells cultured under experimental condition or intervention media containing no cells were physically disconnected from the neurons in the center dish.

We considered the possibility that if physically disconnected neurohormonal independent, non-diffusible cell-cell communication path- ways do exist the presence of a substantial research activity with other cell studies in the area might influence our findings. Therefore, we performed all experiments in a new building with a near empty 30,000 sq. ft. basic science floor that had no other active research activities. In addition, recordings were made in Faraday cage to protect neurons from electromagnetic influence.

The human SH-SY5Y neuroblastoma cells (ATCC CRL-2266) were cultured in a medium consisting of a 1:1 mixture of ATCC-formulated Eagle's Minimum Essential Medium and Ham's F-12 medium containing 10% heat-inactivated FBS, 4mM glutamine, 100 U/mL penicillin, 100 mg/mL streptomycin and 0.25 mg/mL ampho- tericin B in 5% (v/v) CO_2 and balanced moist air at 37°C. Ca2+ fluorescence imaging was carried out as previously described [3]. The coverslips were mounted in a RC-26 recording chamber P-4 (Warner Instruments, Hamden, CT) and placed on a stage of Olympus IX51 inverted microscope (Olympus America, Center Valley, PA). Observations were made at room temperature (20-23°C) with 20X UApo/340 objective. Neurons were bathed and perfused with HBSS buffer using gravity at a rate of 1-2 ml/min. Regions of interest were identified within the soma from which quantitative measurements were made by re-analysis of stored image sequences using Slidebook® Digital Microscopy software. $[Ca^{2+}]_i$ was determined by ratiometric method of Fura-2 fluorescence from calibration of series of buffered Ca^{2+} standards.

EFFECT OF ESTRADIOL ON ATP-INDUCED $[CA^{2+}]_I$ IN SENSORY NEURONS IN THE PRESENCE OF PHYSICALLY ISOLATED LOCAL DRG CELLS

Consistent with prior studies, we observed that a brief 10 second application of ATP (10μM) by fast superfusion produced equal $[Ca^{2+}]_i$ spikes in center and surround DRG neurons. After a 5-min washout with

HBSS, additional stimulation with ATP (10μM) induced sub- sequent $[Ca^{2+}]_i$ transients. Pretreatment with purinergic receptor antagonist PPADS (5 μM) blocked the ATP-induced $[Ca^{2+}]_i$ transients. Similarly, ATP stimulation in a Ca^{2+}-free media in the presence of the Ca^{2+} chelator, BAPTA (10 mM), eliminated $[Ca^{2+}]_i$ spikes indicating the necessity for P2X3 receptors and extracellular Ca^{2+}. 17β-estradiol (E2) (100 nM) by itself had no effect on basal $[Ca^{2+}]_i$ but potently attenuated ATP-induced $[Ca^{2+}]_i$ transients. The effect of E2 was reversible. 5 minute incubation with E2 reduced ATP-induced $[Ca^{2+}]_i$ transient from 425.86 ± 49.5 nM to 171.17 ± 48.9 nM, $p < 0.05$).

The amplitude of $[Ca^{2+}]_i$ response represented the difference between baseline concentration and the transient peak response to drug stimulation. Differences in response to DRG chemo- sensitive receptor modulators were assessed by comparing $[Ca^{2+}]_i$ increases during the first stimulation with the second. All of the data are expressed as the mean ± SEM. Statistical analysis was performed using Statistical Package for the Social Sciences 18.0 (SPSS, Chicago, IL, USA). To assess the significance among different groups, data were analyzed with one-way ANOVA followed by Schéffe post hoc test. A p <0.05 was considered statistically significant.

EFFECT OF 17-B ESTRADIOL INHIBITION OF ATP/CAPSAICIN-INDUCED CALCIUM SIGNALING IN DRG NEURONS IN THE PRESENCE OF LOCAL PHYSICALLY ISOLATED MEDIA

Brief application of TRPV1 agonist capsaicin (3 second 100 nM) by fast superfusion produced $[Ca^{2+}]_i$ spikes which were almost completely blocked by 100 nM capsazepine, a TRPV1- selective antagonist. Since the effect of capsaicin was non-reversible we applied the estradiol first assuming its non-reversibility on TRPV1 receptors. The E2 by itself had no effect on basal $[Ca^{2+}]_i$, but E2 (100 nM) attenuated the peak of capsaicin-induced $[Ca^{2+}]_i$ transients from 399.21 ± 44.5 nM to 175.01 ± 34.8 nM (Figure 4. 2).

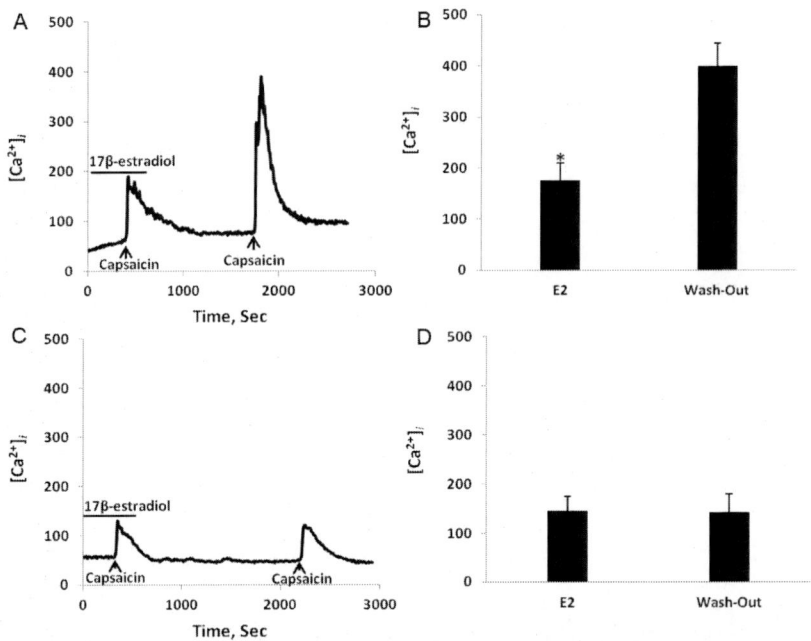

Figure 4.2. Effect of E2 on capsaicin-induced $[Ca^{2+}]_i$ increase in control and in the local presence of physi- cally disconnected SH-SY5Y cells. A: Pre-incubation for 5 min with 17β-estradiol (100 nM) attenuates capsaicin (300 nM)- induced $[Ca^{2+}]_i$ response (first stimulation) compared with controlled capsaicin response (second stimulation). B: Summary of capsaicin- induced $[Ca^{2+}]_i$ increase in the presence of E2 and after wash-out with experimental medium. Values are expressed as mean ± SEM (n=8). * indicates statistically significant difference from control, $P<0.05$. C: Effect of 17β-estradiol (100 nM) on capsaicin (300 nM)-induced $[Ca^{2+}]_i$ increase was abolished after 12 hours incubation near but physically-disconnected from human neuroblastoma SH-SY5Y cells. D: Summary of capsaicin- induced $[Ca^{2+}]_i$ increase in the presence of E2 and after wash-out with experimental medium in sensory neurons cultured in the local presence of physically-disconnected SH-SY5Y cells. Values are expressed as mean ± SEM (n=4).

We compared ATP/capsaicin-induced $[Ca^{2+}]_i$ spikes from DRG neurons in the inner "center" culture dish in the presence of only DRG media or SH-SY5Y media in the outer "surround" cul- ture dish (see Material and Methods). In these experimental settings we did not detect any sig- nificant difference in DRG neuron response to ATP (10 µM), capsaicin (100 nM) or 17β-estradiol (100 nM) application compared to control set- ting 1 conditions above with normal DRG in the outer "surround" culture dish.

EFFECT OF ESTRADIOL INHIBITION OF ATP/CAPSAICIN-INDUCED CALCIUM SIGNALING IN DRG NEURONS IN THE PRESENCE OF LOCAL PHYSICALLY ISOLATED SH-SY5Y NEUROBLASTOMA CELLS

We compared ATP/capsaicin-induced $[Ca^{2+}]_i$ spikes from inner "center" culture dish DRG in the presence of neuroblastoma SH-SY5Y cells cultured in the outer "surround" culture dish (see Material and Methods). Both ATP (10 µM) and capsaicin (100 nM)- induced $[Ca^{2+}]_i$ responses were significantly reduced (2 fold; p<0.05) compared to dish-in-dish cultures where DRG cells occupied the surround (193.45 ± 40.16 nM, and 145.90 ± 28.84 nM respectively, vs. 425.86 ± 49.5 nM, and 399.21 ± 44.5 nM in control setting. The effect of E2 on ATP-induced P2X3-mediated $[Ca^{2+}]_i$ observed under control conditions was abolished with neuroblastoma SH-SY5Y cells in the surround. Moreover, E2 similarly abolished capsaicin-induced TRPV1- mediated $[Ca^{2+}]_i$ flux.

EFFECT OF ATP/CAPSAICIN-INDUCED CALCIUM SIGNALING IN DRG NEURONS IN THE PRESENCE OF LOCAL PHYSICALLY ISOLATED DRG NEURONS EXPOSED TO KCl (APOPTOSIS INTERVENTION)

In another set of experiments, we compared ATP-induced $[Ca^{2+}]_i$ spikes from inner "center" culture dish DRG in the presence of DRG neurons exposed to KCl (50 mM >80% apoptosis 12 hours following KCl exposure) in the outer "surround" culture dish. In control, ATP (10 μM) induced equal $[Ca^{2+}]_i$ responses 387.63 ±17.3 nM but failed to produce a typical $[Ca^{2+}]_i$ influx in DRG cells incubated in the local presence of physically-separated dying DRG cells. There was a 4 fold increase in $[Ca^{2+}]_i$ concentration during the first exposure to ATP stimulation (p<0.05), followed by no increase in $[Ca^{2+}]_i$ with further ATP stimulation. Brief application of TRPV1 agonist capsaicin (100 nM) by fast superfusion produced equal $[Ca^{2+}]_i$ spikes 355.83 ± 57.23 nM.

We were able to compare, in a controlled way, the influence of the outer chamber surround cell population on inner chamber DRG calcium signaling in response to well described DRG chemosensitive receptor stimuli. Direct application of ATP and capsaicin to activate DRG cells in the inner chamber in control settings demonstrated increased $[Ca^{2+}]_i$ in DRG in response to ATP as well as capsaicin in the con- trol setting and attenuation of ATP and capsaicin increases in $[Ca^{2+}]_i$ by 17β-estradiol consistent with prior studies [4].

By contrast, the local presence of cancer cells or apoptotic cells in the outer chamber significantly altered inner chamber DRG $[Ca^{2+}]_i$ responses to ATP and capsaicin, as well as 17β-estradiol in different ways.

In the local presence of cancer cells ATP andattenuated and the effect of 17β-estradiol was abolished. The same media minus the cells however, did not alter inner chamber DRG $[Ca^{2+}]_i$ responses indicating the local presence of cells rather than the media was responsible for this effect. In the local presence of apoptotic DRG cells a different response was noted. The initial ATP stimulated increase in $[Ca^{2+}]_i$ was dramatically exaggerated, while subsequent exposure to ATP led to no response in $[Ca^{2+}]_i$. In addition, neither initial nor subsequent exposure to capsaicin induced any response in inner chamber DRG $[Ca^{2+}]_i$. The exaggerated increase in $[Ca^{2+}]_i$ concentration during the first exposure to ATP stimulation in normal DRG nearby DRG cells undergoing apoptosis is similar to that seen in murine neurons following sodium fluoride toxicity which was reported to increase intercellular Ca^{2+} concentration leading to apoptosis [5]. While further studies are needed to better understand the observed $[Ca^{2+}]_i$ changes and their implications. Our findings demonstrate that apoptotic and cancerous cells are capable of exerting a non-diffusible, non-neuronal influence over distance on nearby, but physically disconnected cells through Ca^{2+} signaling.

Our findings suggest non-visceral nociception and nociceptor sensitization may also be modulated by P2X3 and capsaicin. On the other hand, this study does not identify the mechanism(s) of action by which the physically disconnected, non-diffusible cell-to-cell signaling changes observed in DRG receptor mediated $[Ca^{2+}]_i$ fluxes are mediated, e.g., gene expression, epigenetic modification or other. Changes in $[Ca^{2+}]_i$ fluxes could be mediated through modification of ion and calcium related gene expression (i.e., g proteins) as reported by Xiao et al. who identified marked changes following DRG axotomy with respect to the expression of over 170 genes including neuropeptides, receptors, ion channels, and signal transduction molecules [6].

Limitations of our study are that although the inner chamber is enclosed we still cannot exclude the possibility of volatile communication via aromatic compounds. Activation of phospholipase can potentially serve as biochemical messengers across nearby cellular colonies. In addition, other potent signal transduction gases including NO and CO produced by nNOS and Heme oxygenases can be carried in air from one compartment to other where through formation of second messenger (e.g., cGMP) can alter $[Ca^{2+}]_i$ and cell function in the cells residing in the neighboring compartment. Finally, hydrogen sulfide (H_2S) is another recognized signal transduction gas produced by cystathionine c-lyase or cystathionase, cystathionine b-synthase and 3-mercaptopyruvate sulphurtransferase. H_2S activates ATP-sensitive potassium channels and large-conductance Ca^{2+}-activated potassium channels, and can induce changes in mitogen-activated protein kinase, cell cycle-related kinase, cell death- related gene and ion channels [7].

In addition, we did not explore a dose/distance relationship between the inner and outer chambers, or vary the number of local cancer or apoptotic cells. Also, we only looked at the effects of two different cell types/conditions on the inner DRG cells and we did not examine other cell types in the inner chamber. However, we felt these additional permutations while important were beyond the scope of this initial study, which was to examine if physically distinct non- diffusible cell-cell communication could be identified.

Local presence of human neuroblastoma SH-SY5Y cells or DRG undergoing KCl induced apoptosis alter both the direct and 17β-estradiol regulated effect of ATP and capsaicin induced P2X3 and TRPV1 receptor mediated $[Ca^{2+}]_i$ in mouse DRG neurons through what appears to be a non-local form of cell-to-cell communication. The inner enclosure suggests aerosolized communication of aromatic hydrocarbons is unlikely, though not absolutely excluded. Further studies are needed to confirm our findings and to explore the exact mechanism of this action on calcium signaling.

REFERENCES

[1] Lenfant C. Shattuck lecture--clinical research to clinical practice--lost in translation? *New England Journal of Medicine.* 2003; 349: 868-874.

[2] Norris K. Translational research: moving scientific advances into the real world setting. *Ethnicity & Disease* 2005; 15: 363-364.

[3] Chaban V., Mayer E., Ennes H. and Micevych P. Estradiol inhibits ATP-induced intracellular calcium concentration increase in dorsal root ganglia neurons. *Neuroscience* 2003; 118: 941-948.

[4] Cho T. and Chaban V. Interaction between P2X3 and oestrogen receptor ERα/ERβ in ATP- mediated calcium signalling in mice sensory neurones. *Journal of Neuroendocrinology* 2012; 24: 789-797.

[5] Haojun Z, Yaoling W, Ke Z, Jin L and Junling W. Effects of NaF on the expression of intracellular Ca^{2+} fluxes and apoptosis and the antagonism of taurine in murine neuron. *Toxicology Methods* 2012; 22: 305-308.

[6] Xiao H., Huang Q., Zhang F., Bao L., Lu Y., Guo C., Yang L., Huang W., Fu G., Xu S., Cheng X., Yan Q., Zhu Z., Zhang X., Chen Z., Han Z. and Zhang X. *Identification of gene expression profile of dorsal*

root ganglion in the rat peripheral axotomy model of neuropathic pain. Proceedings of the National Academy of Sciences USA 2002; 99: 8360-8365.

[7] Yang GD and Wang R. H$_2$S and cellular proliferation and apoptosis. *Sheng Li Xue Bao* 2007; 59: 133-140.

Chapter 5

CALCIUM SIGNALING AND SENSORY NEURONAL REORGANIZATION

ABSTRACT

Modification in calcium signaling could be one of the possible explanations for the phenomenon of neuronal reorganization associated with pain transmission (nociception). Moreover, clinical presentations of functional syndromes often lack a specific pathology in the affected organ but may respond to a visceral cross-sensitization using Ca^{2+} as a second messenger in which increased nociceptive input from an inflamed organ sensitizes neurons that receive convergent input from an unaffected organ.

ROLE OF SENSORY NEURONS IN NOCICEPTION

Spinal dorsal DRGs are an attractive therapeutic target for alleviating pain as they serve as a gate for the transmission of information prior to the central nervous system. Pain signaling is regulated within DRG before reaching the central nervous system (CNS), making these primary afferents an important target for therapy.

Spinal interneurons integrate nociceptive sensory inputs from dorsal root ganglia to the dorsal horn and gate their access to spinal projection neurons, as was described by Ronald Melzack more than 50 years ago in the Gate Control Theory of Pain Processing [1]. Today, it is evidenced that dorsal horn interneurons limit access of non-nociceptive sensory input to nociceptive-specific projections in the spinal cord [2]. Chronic pain leads to hypersensitivity through convergence of nociceptive and non-nociceptive inputs on spinal projection neurons that signal to the primary somatosensory cortex.

Visceral sensitization is an important underlying contributor to chronic pain states but remains poorly understood. The mechanisms underlying how conduction is gated by depolarization or hyperpolarization of nociceptors and overall pain information through the DRG is of great interest of scientific and medical community. In this mini-review we present new insights on the cellular bases of visceral pain signaling in the peripheral and central nervous system with potential to understand the complex neuronal regulation and reorganization of the nociceptive network.

Differences between C, Aδ, and Ab fibers, heterogeneity of channel expression significantly affect the DRG properties and modulation of their conduction. Population of neurons affected by voltage- and ligand-channel activation may be sufficient to influence behavioral response to nociceptive stimuli. Depolarization can trigger depolarization block and inhibition of pain signaling in substantial number of neurons within the same ganglion. Indeed, many channelopathies are associated with neuropathic pain [3].

Visceral sensitization is an important underlying contributor to chronic pain states but remains poorly understood. The mechanisms underlying how conduction is gated by depolarization or hyperpolarization of nociceptors and overall pain information through the DRG is of great interest of scientific and medical community.

PERIPHERAL MODULATION OF SENSORY NEURONS EXCITABILITY

Several lines of evidence indicate that estrogen-mediated calcium signaling directly influence the functions of primary afferent neurons. Both subtypes of estrogen receptors (ERα and ERβ) are present in DRG neurons including the small- diameter putative nociceptors [4]. *In vitro,* ATP-sensitive DRG neurons respond to E2, which correlated well with the idea that visceral afferents are E2 sensitive: i) visceral pain is affected by hormonal level in cycling females [5]; ii) there are sex differences in the prevalence of functional disorders involving the viscera [6]; and iii) putative visceral afferents fit into the population of DRG neurons that are sensitive to E2. These data suggest that in addition to CNS actions of E2 [7], E2 can act in the periphery to modulate nociception.

The inflammatory process produces mediators which activate nociceptors by interacting with ligand-gated ion channels or by sensitizing primary afferents [8]. One mechanism for sensitization involves phosphorylation of ion channels and receptors including P2X3 and TRPV1 receptors. Inflammation does not change the percentage of total cells responding to ATP but sensitizes the ATP response by increasing the expression of P2X3. Thus, the greater behavioral sensitivity during the inflammation is due to a twofold to threefold increase in ATP responses suggesting that a small amount of ATP would evoke depolarization sufficient to elicit action potentials in DRG neurons. This pathological response arises from sensitization of DRG to external stimuli. Furthermore, inflamed tissues augment nociceptor responsiveness by acting on TRPV1 [9]. Gastrointestinal inflammation modulates the intrinsic properties of nociceptive dorsal root ganglia neurons, which innervate the GI tract and these changes are important in the genesis of abdominal pain and visceral hyperalgesia. Neurons exhibit hyperexcitability characterized by a decreased threshold for activation and increased firing rate [10].

Sensitization may also account for a lowered nociceptive threshold to mechanical manipulation of the inflamed area. Within the context of our cross-sensitization hypothesis, inflammation sensitizes non-inflamed viscera that are innervated by the same DRG and/or cross-sensitization occurs as a result of intra-DRG release of sensitizing mediators such as ATP within the DRG.

ERα/ERβ receptors were traditionally envisioned as E2-activated transcription factors. However, more recent studies indicate that E2 has a multiplicity of actions: membrane, cytoplasmic and nuclear. E2 modulates cellular activity by altering ion channel opening, G-protein signaling, and activation of trophic factor-like signal transduction pathways. These effects have been ascribed to membrane-associated receptors. Moreover, estrogen receptors (ERs) are distributed in regions of the central and peripheral nervous system that mediate nociception. Results from our laboratory and others indicate that E2 acts in DRG neurons to modulate L-type voltage-gated calcium channels (VGCC) and through group II metabotropic glutamate receptors [11]. E2 has a significant role in modulating visceral sensitivity, indicating that E2 alterations in sensory processing may underlie sex-based differences in pain symptoms (Figure 5.1).

Little is known about E2-mediated mechanisms in peripheral nervous system, but the fact that DRG neurons express ERs and respond to E2 treatment suggest that they are a potential target for calcium changes associated with nociception.

From a public health prospective, the outcome of this study will have a substantial impact on our knowledge of nociceptive functional diseases and help achieve a deeper understanding of gender differences presented in clinical aspects of these symptoms. Only a thorough understanding of the mechanism implicated in these phenomena can truly contribute to the designing of new and more efficient therapies.

E2 modulates nociceptive responses in pelvic pain syndromes; however, this mechanism remains unresolved. Within the context of our hypothesis E2 modulation of nociceptive response depends on the type of pain, its durations and the involvement of other nociceptive-mediated mechanisms. Visceral nociception appears to be regulated by ATP/capsaicin and E2 in the periphery at the level of DRG. The DRG is an important site of visceral afferent convergence and cross-sensitization.

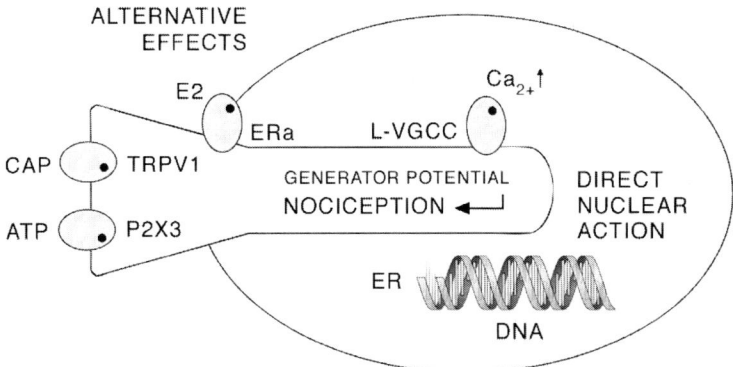

Figure 5.1. Proposed mechanism of 17-βestradiol (E2) effect on ATP-induced $[Ca^{2+}]_i$ signaling in primary sensory neurons. ATP released by tissue damage acts on P2X3 receptor resulting in activation of the L-type voltage-gated calcium channel (VGCC). ERα modulates ATP-induced P2X3- mediated as well as capsaicin-induced TRPV1- mediated $[Ca^{2+}]_i$ increases [12].

REFERENCES

[1] Melzack, R. and Wall P. Pain mechanisms: a new theory. *Science,* 1965, 150: p. 971- 979.

[2] Peirs, C., Williams S., Zhao X., Walsh C., Gedeon J., Cagle N., Goldring A., Hioki H., Liu Z, Marell P., Seal R. *Dorsal Horn Circuits for Persistent Mechanical Pain. Neuron,* 2015, 87(4): p. 797-812.

[3] Smith E. Advances in understanding nociception and neuropathic pain. *Journal of Neurology,* 2018; 265(2): 231–238.

[4] Papka, R. and Storey-Workley, M., *Neuroscience Letters,* 2002; 319: 71- 74.

[5] Traub, R. J. and Murphy, A. 2002, *Pain,* 95: 93-102.

[6] Sand, P. *Journal of Reproductive Medicine,* 2004, 49: 230- 234.

[7] Aloisi, A. M., Ceccarelli, I. and Herdegen, T. *Neuroscience Letters,* 2000, 281: 29- 32.

[8] Dai, Y., Fukuoka, T., Wang, H., Yamanaka, H., Obata, K., Tokunaga, A. and Noguchi, K., *Pain,* 2004, 108: 258-66.

[9] Negri, L., Lattanzi, R., Giannini, E., Colucci, M., Margheriti, F., Melchiorri, P., Vellani, V., Tian, H., De Felice, M. and Porreca, F., *Journal of Neuroscience*, 2006, 26: 6716- 6727.
[10] Beyak, M. J. and Vanner, S. *Neurogastroenterology and Motility*, 2005, 17: 175-186.
[11] Chaban, V., Li, J., McDonald, J., Rapkin, A. and Micevych, P., *Journal of Neuroscience Research*, 2011, 89: 1707- 1710.
[12] Chaban V. Estrogen modulation of visceral nociceptors. *Current Trends in Neurology*, 2014, 7: 51- 56.

Chapter 6

ROLE OF TRPV1 AND P2X2/3 RECEPTORS IN CA^{2+}– MEDIATED NOCICEPTION IN VISCERAL SENSORY NEURONS

ABSTRACT

One of possible mechanism of Ca^{2+}- mediated nociception may be the convergence of nociceptive stimuli on the primary sensory neurons which innervate viscera. E2 shows preferentially on visceral afferents to modulate the transmission of visceral pain. In this study we used visceral and cutaneous afferent neurons and compared E2 modulation of P2X2/3 and TRPV1 mediated calcium changes in cultured retrogradely- labeled DRG neurons. E2 significantly decreased the α,β-me ATP-/capsaicin-induced $[Ca^{2+}]_i$ transients in C57Bl/6J (wild type) and ERβKO (ER-α knock- out) mice, but this hormone had no effect in ERαKO mice indicating that this situation depends on membrane associated ERα receptors. In addition, Prostaglandin E2 potentiated capsaicin-induced calcium fluxes showed 2~3 fold amplitudes with prolonged duration of the response. The effect of this inflammatory mediator also was estrogen receptor alpha-dependent. However, there are no effect and smaller E2 effect with/without PGE2 on α,β-meATP/capsaicin-induced $[Ca^{2+}]_i$ responses in retrogradely- labeled somatic DRG neurons compared with visceral DRG neurons. Together our data suggest that a novel communication between P2X2/3 or TRPV1 and membrane-associated ER-α in primary visceral sensory neurons and a novel mechanism to understand sex differences observed in clinical studies of many functional diseases related with visceral nociception.

ESTRADIOL MODULATION OF ATP-INDUCED CALCIUM FLUXES DIFFERENTLY REGULATES VISCERAL AND CUTANEOUS NOCICEPTIVE SIGNALING IN MOUSE DRG NEURONS

Animals

We have used 6~8 week old female wild type (Wt, C57Bl/6J), ERαKO (B6.129P2-Esr1^{tm1Ksk}/J), and ERβKO (B6.129P2-Esr2^{tm1Unc}/J) mice (Jackson Laboratory, Bar Harbor, ME). Upon arrival mice were housed in microisolator caging and maintained on a 12h light/dark cycle in a temperature-controlled environment with access to food and water ad libitum for two weeks. In some experiments we used animals from our breeding colony. All studies were carried out in accordance with the guidelines of U.S. National Institute of Health (NIH) Guide for the Care and Use of Laboratory Animals.

Primary Culture of DRG Neurons

The isolation procedure and primary culture of mouse lumbosacral DRG has been published in detail. DRG tissues were obtained from C57Bl/6J (30 g), ERαKO and ERβKO (Jackson Laboratory; 20 g) transgenic types. Briefly, lumbosacral adult DRGs (level L1-S1) were collected under sterile technique and placed in ice-cold medium Dulbecco's Modified Eagle's Medium (DMEM; Sigma-Aldrich St. Louis, MO). Adhering fat and connective tissue were removed and each DRG was minced with scissors and placed immediately in a medium consisting of 5 ml of DMEM containing 0.5 mg/ml of trypsin (Sigma, type III), 1 mg/ml of collagenase (Sigma, type IA) and 0.1 mg/ml of DNase (Sigma, type III) and kept at 37°C for 30 minutes with agitation. After dissociation of the cell ganglia, soybean trypsin inhibitor (Sigma, type III) was used to terminate cell dissociation. Cell suspensions were centrifuged for one minute at 1000 rpm and the cell pellet were resuspended in DMEM supplemented with 5% fetal bovine serum, 2 mM glutamine-penicillin-streptomycin mixture, 1 µg/ml DNAase and 5 ng/ml NGF (Sigma). Cells were placed on Matrigel® (Invitrogen, Carlsbad, CA)- coated 15-mm coverslips (Collaborative Research Co., Bedford, PA) and kept at 37°C in 5% CO_2 incubator for 24h, given fresh media and maintained in primary culture until used for experimental procedures.

Retrograde Labeling

Retrograde labeling was used to confirm DRG neurons that had innervated both visceral organs: uterus and colon. All surgical protocols were performed under sterile environments within the designated animal surgery room. For colon afferents, the descending colon of female mice (Wild type, ERαKO, and ERβKO) was exposed under anesthesia with isoflurane. Fluorogold (FG, 1.5% in .9% saline) was injected into the intestinal muscle wall into five to six different sites (1 - 2 µl / site, total of 10 µl / mice) by using a Hamilton syringe (Hamilton Co., Reno NV) with a 31-gauge needle. In another experiments, we used uterus-specific DRG neurons in which tetramethylrhodamine (TMR, 1.0% in .9% saline) tracer was injected into the uterus (total of 20 µl/mice). Injection sites were carefully swabbed and the colon and uterus was extensively rinsed with 0.9% sodium chloride solution. The abdomen was sutured and the animals monitored for signs of pain or discomfort during survival period. Both dyes are injected into respective place into colon and uterus of the same animals. Immediately after surgery, mouse were injected with sterile saline to prevent dehydration and were given analgesics for pain relief and antibiotics. For somatic labeling, 50 µl cholera toxin conjugated to Alexa Fluor 647 (2.5% solution in PBS; Molecular Probes, Eugene, OR) was injected subcutaneously into five sites at the mouse hind paw. All mouse were allowed to survive 7 - 10 days to allow for maximal transport of retrograde neuron tracer and housed in groups of two under 12/12 hours light cycle with food and water available ad libitum.

$[Ca^{2+}]_i$ Fluorescence Imaging

Ca^{2+} fluorescence imaging was carried out as previously described [23-24]. DRG neurons were loaded with fluorescent dye 5 mM Fura-2 AM (Invitrogen, Carlsbad, CA) for 45 min at 37°C in HBSS supplemented with 20 mM HEPES, pH 7.4. The coverslips were mounted in a RC-26 recording chamber P-4 (Warner Instruments, Hamden, CT) and placed on a stage of Olympus IX51 inverted microscope (Olympus America, Center Valley, PA). Observations were made at room temperature (20-23°C) with 20X UApo/340 objective. Neurons were bathed and perfused with HBSS buffer using with using gravity at a rate of 1-2 ml/min. Fluorescence intensity at 505 nm with excitation at 334 nm and 380 nm were captured as digital images (sampling rates of 0.1- 2s). Regions of interest were identified within the soma from which quantitative measurements were made by re-analysis of stored image sequences using Slidebook® Digital Microscopy software. $[Ca^{2+}]_i$ was determined by ratiometric method of Fura-2 fluorescence from calibration of series of buffered Ca^{2+} standards. E2 was applied acutely for five minutes onto the experimental chamber. Repeated application of drugs was achieved by superfusion in a rapid mixing chamber into individual neurons for specific intervals (100-500 ms). We calculated actual $[Ca^{2+}]_i$ in areas of interest in each neurons with the formula:

$$[Ca^{2+}]_i = K_d \times (R-R_{min})/(R_{max}-R) \times \beta$$

where K_d is the indicator's dissociation constant of the fluoroprobe; R is ratio of fluorescence intensity at two different wavelengths (340/380 nm for fura-2); R_{max} and R_{min} are the ratio at fura-2 with an saturated Ca^{2+} and free Ca^{2+}. β is the ratio of the denominators of the minimum and maximum conditions.

Statistical Analysis

The amplitude of $[Ca^{2+}]_i$ response represents the difference between baseline concentration and the transient peak response to drug stimulation. Significant differences in response to chemical stimulation were obtained by comparing $[Ca^{2+}]_i$ increases during the first stimulation with the second. All of the data are expressed as the mean ± SEM. Statistical analysis was performed using Statistical Package for the Social Sciences 18.0 (SPSS, Chicago, IL, USA). To assess the significance among different groups, data were analyzed with one-way ANOVA followed by Schéffe post hoc test. A $p < 0.05$ was considered statistically significant.

DRG neurons innervating viscera are about 5-10%. A corollary of that hypothesis is that cutaneous pain may show different action of pain modulation compared with visceral pain. Even though there are no changes in cutaneous modulation of pain, pathologically visceral modulation of pain such as inflammation and IBS is more prevalent in women [1]. This experiment is designed to compare the pharmacology of retrogradely labeled visceral DRG neurons with retrogradely labeled cutaneous DRG neurons. Within the context of our hypothesis, nociceptive responses by E2 modulation depend on the kinds of pain, its durations.

To find mechanism of viscero/visceral hyperalgesia between organs including viscero/visceral-somatic convergent neurons, retrograde labeling was used to demonstrate cultured DRG neurons receiving sensory signaling input from different visceral organs such as colon and uterus. We inject fluorogold (FG) to colon and tetramethylrhodamine (TMR) to uterus into the muscle wall by abdominal incision and somatic afferent neurons (Alexa Fluor 647) by subcutaneous injection into mouse hind paw. Both uterine and colonic afferents were shown in DRG neurons.

Retrogradely- labeled visceral DRG cells that project to the colon or uterus were co-localized in the same DRG neuron, suggesting a subpopulation of DRG neurons that innervate both colon and uterus. These DRG neurons innervating both colon and uterus represent a new group of dichotomizing sensory afferents. In addition, E2 (100 nM) by itself had no effect on basal $[Ca^{2+}]_i$, but the ATP-induced $[Ca^{2+}]_i$ flux was decreased by this hormone. After the initial ATP response, incubation with E2 for five minute inhibited ATP-induced $[Ca^{2+}]_i$ transient. On the other hand, retrogradely- labeled cutaneous DRG neurons injecting to hind paw were shown in the DRG cells. These neurons did not block ATP-induced $[Ca^{2+}]_i$ and showed low amplitudes of E2 effects on ATP-induced $[Ca^{2+}]_i$ response.

THE EFFECT OF E2 MODULATION OF P2X2/3 AND TRPV1 RECEPTORS MEDIATED $[CA^{2+}]_I$ RESPONSE IN BOTH RETROGRADELY-LABELED VISCERAL AND CUTANEOUS DRG NEURONS

DRG neurons innervating viscera demonstrated greater amplitudes of E2 effects on ATP-induced $[Ca^{2+}]_i$ response suggesting that neurons innervating visceral organs may express and regulate ER differently. ER is more tightly coupled to P2X2/3 and TRPV1. To identify of E2 modulation of calcium responses from retrogradely labeled visceral and somatic neurons, we also used α,β-meATP, a specific agonist of P2X2/3, to confirm the observations presented with ATP (n=69 cells/each group 7 mouse). Both P2X2/3 and P2X3 receptors have similar properties and seem to be a target for estradiol (Figure 6.1a) 382.60 ± 40.09 vs. 216.84 ± 38.64 nM, n=63 cells/each group 7 mouse, $p<0.05$. Moreover, TRPV1 receptors can be activated by capsaicin and regulated by inflammatory mediators including PGE2 and bradykinin [2], whose effects are mediated through an increase of $[Ca^{2+}]_i$ in mouse DRG neurons. The E2 had no effect of calcium changes by itself, but E2 (100 nM) showed low intensity of capsaicin-induced $[Ca^{2+}]_i$ transients (Figure 6.1b). The superfusion with capsaicin (100 nM) induced $[Ca^{2+}]_i$ increases to 354.33 ± 41.11 nM, but this effect was diminished to 199.81 ± 44.18 nM by E2. There was statistically significant difference the effect of TRPV1 in wild type mice. However, there are no effect and smaller E2 effect on α,β-meATP/capsaicin-induced $[Ca^{2+}]_i$ responses in retrogradely labeled somatic DRG neurons compared with visceral DRG neurons (Figure 6.1c and 6.1d).

Our results suggest that the E2 effects on TRPV1 seemed to be one of the signaling pathways for E2-mediated changes of TRPV1 activity and the pain perception of visceral afferents.

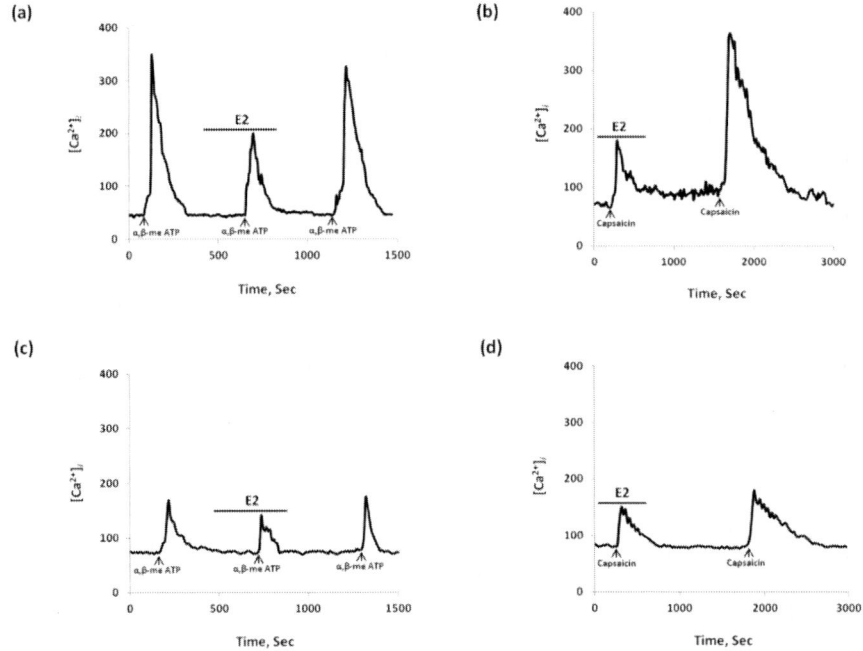

Figure 6.1. E2 effects of α,β-me ATP-/capsaicin-induced $[Ca^{2+}]_i$ transients in visceral and cutaneous sensory DRG neurons from wild type mice. (a) α,β-me ATP-induced $[Ca^{2+}]_i$ response rapidly attenuated by E2 (100 nM) in visceral DRG neurons. (b) The first capsaicin-induced $[Ca^{2+}]_i$ response also rapidly decreased by E2 on visceral DRG neurons. E2 didn't inhibit (c) α,β-me ATP/ (d)capsaicin-induced $[Ca^{2+}]_i$ response and this showed low intensities on α,β-me ATP/capsaicin-induced $[Ca^{2+}]_i$ response.

The Role of ERa/ERb in E2 Modulation on a,b-meATP- and Capsaicin-Induced $[Ca^{2+}]_i$ Transients in Retrogradely-Labeled Visceral and Cutaneous Mouse DRG Neurons

To verify the role of ERα and ERβ in the potentially different response from visceral and somatic DRG neurons, visceral DRG neurons retrogradely- labeled with FG for colonic afferents and TMR for uterine afferents were stimulated with either α,β-meATP or capsaicin and compared to differences from retrogradely labeled somatic DRG neurons in knock-out mice. The E2 effects in ERβKO mouse DRG neurons show similar to wild type mice (Figure 6. 2b). However, in ERαKO mice there are no E2 effects on α,β-meATP-induced $[Ca^{2+}]_i$ fluxes (Figure 6. 2a). 17α-estradiol, a different isomer of 17β-estradiol, had no effect on Wt, ERαKO or ERβKO mice.

We also demonstrated capsaicin-induced $[Ca^{2+}]_i$ changes in knock-out mice to find the examination presented with capsaicin in wild type mice. We also found the similarity of E2 effects in ERβKO to that presented in wild type mice (Figure6.3b) but E2 didn't show any differences in ERαKO mice on capsaicin-induced $[Ca^{2+}]_i$ fluxes (Figure 6.3a). Our study used a combined approach to decide that primary cultured DRG neurons can be used to research the cellular reaction to a nociceptive α,β-meATP-sensitive P2X2/3 and capsaicin-sensitive TRPV1 receptors. In retrogradely labeled somatic DRG neurons, however E2 didn't inhibit and show low intensity on α,β-meATP (Figure 6.3c and 6.3d), capsaicin-induced $[Ca^{2+}]_i$ fluxes (Figure 6. 4c and 6.4 d).

Our data suggest that sensory DRG neurons innervating viscera demonstrate greater amplitudes of E2 effects on the α,β-meATP/capsaicin-induced $[Ca^{2+}]_i$ fluxes and E2 interfere with the membrane-associated ERα in viscerally-specific neurons.

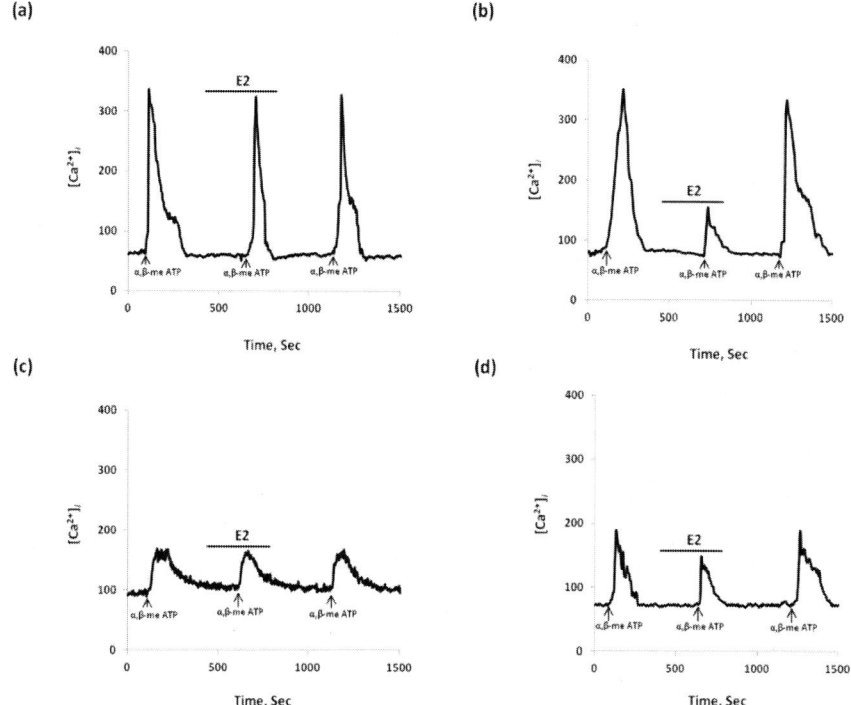

Figure 6.2. The effect of E2 on α,β-me ATP-induced $[Ca^{2+}]_i$ transients in visceral and somatic DRG neurons from estrogen receptor-α knockout (ERαKO) and estrogen receptor-β knockout (ERβKO) mice. (a) In ERαKO mouse, E2 added for 5 min didn't inhibit α,β-me ATP-induced $[Ca^{2+}]_i$ transient in visceral DRG sensory neurons; (b) In ERβKO mouse E2 stimulation significantly attenuated the α,β-me ATP-stimulated $[Ca^{2+}]_i$ transient similar to that observed in visceral DRG neurons from Wt mice. E2 didn't decrease and somatic DRG neurons showed low amplitudes on α,β-me ATP-induced $[Ca^{2+}]_i$ in (c) ERαKO and (d) ERβKO mice.

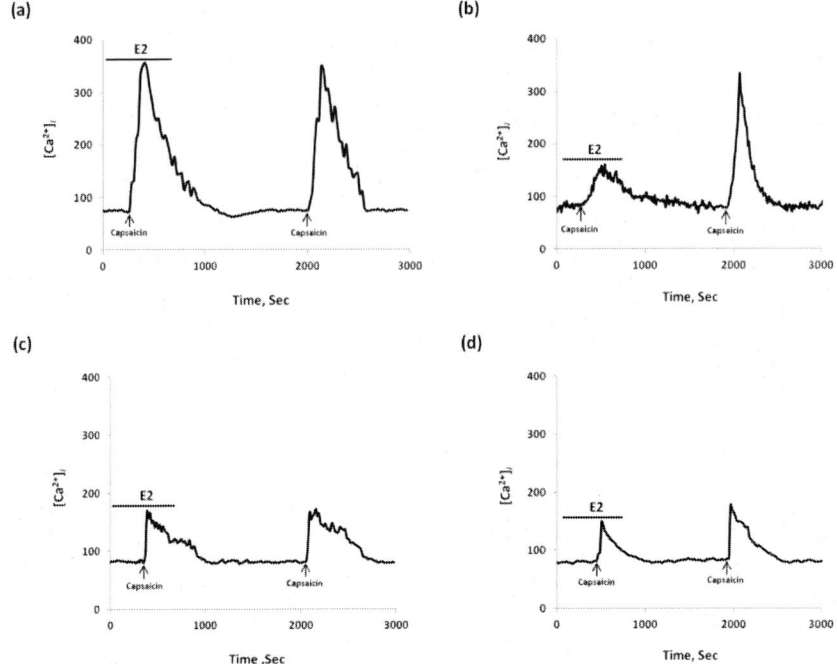

Figure 6.4. The E2 effects with visceral and somatic DRG neurons on capsaicin-induced $[Ca^{2+}]_i$ fluxes in ERαKO and ERβKO mice. (a) E2 didn't inhibit capsaicin-induced $[Ca^{2+}]_i$ fluxes in visceral DRG sensory neurons from ERαKO mice. (b) E2 stimulation significantly attenuated the capsaicin-stimulated $[Ca^{2+}]_i$ fluxes similar to that observed in visceral DRG neurons from wild type mice. In somatic DRG neurons, there is no significant attenuation and low amplitudes on capsaicin-induced $[Ca^{2+}]_i$ in (c) ERαKO and (d) ERβKO mice compared with visceral DRG neurons.

COMPARISON OF THE EFFECTS OF ESTRADIOL ON PROSTAGLANDIN PGE$_2$ POTENTIATION OF TRPV1-MEDIATED CALCIUM RESPONSE IN RETROGRADELY-LABELED VISCERAL AND CUTANEOUS DRG NEURONS FROM WT, ERαKO, AND ERβKO MICE

The action of E2 is important to determine the physiological properties of ERs expressed in visceral neurons and which ER mediates E2 effects in visceral compared with somatic DRG neurons. In addition, methodologically it is very difficult to study receptor activation on the neuritis of DRG neurons. An underlying assumption of these studies is that receptors located on peripheral terminals of colonic primary afferent neurons have the same functional properties as those expressed on the soma of DRG neurons [3]. Hence in vitro DRG neurons are the best model for the proposed experiments. It is hypothesized that ERs on colonic primary afferent neurons may have different functional properties compared with those expressed on somatic afferent neurons. The potentiating of TRPV1 is considered to be important in pain process from inflammation. Moreover, E2 is more likely to magnify higher intensities of PGE2-mediated TRPV1 activation in retrogradely- labeled visceral DRG neurons than in somatic DRG neurons.

To confirm whether ERα or ERβ play a important role in E2 pretreatment before the addition of PGE2 (100 nM) on capsaicin-induced calcium responses, retrogradely- labeled DRG neurons from Wt, ERαKO, and ERβKO mice were used. In retrogradely- labeled visceral DRG neurons, after pretreatment with PGE2, the E2 mediated $[Ca^{2+}]_i$ transients by capsaicin was elevated from a control condition (data not shown). This study suggested that capsaicin-induced $[Ca^{2+}]_i$ transients were enhanced after the PGE2 pretreatment by almost 2~3 fold. However, the E2 by itself had no effect on basal $[Ca^{2+}]_i$, but E2 (100 nM) in the presence of inflammatory mediator, PGE2, attenuated the intensities of capsaicin-induced $[Ca^{2+}]_i$ transients in wild type (Figure 6.5a) 356.83 ± 42.38 nM, n=67 cells/7 mice, $p<0.05$.

We tested the role of ERβ in the context of the influence of PGE2-mediated capsaicin-induced $[Ca^{2+}]_i$ fluxes. The effect of PGE2 after E2 pretreatment was similar in ERβKO mouse DRG neurons to that observed in Wt mice (Figure 6.5c), 339.07 ± 54.83 nM, n=58 cells/7 mice, $p<0.05$) but, E2 did not block capsaicin-induced $[Ca^{2+}]_i$ in DRG neurons from ERαKO mice suggesting that its diminution depends on ERα (Figure 6.5b), 260.12 ± 39.42 nM in control to 583.52 ± 47.52 nM, n=72 cells/7 mice, $p<0.05$). However E2 didn't work on capsaicin-induced $[Ca^{2+}]_i$ responses in the retrogradely- labeled somatic DRG neurons (Figure 6.5b).

Our data suggest that visceral DRG neurons show higher intensities of E2 effects on PGE2-mediated capsaicin-induced $[Ca^{2+}]_i$ fluxes and E2 interfere with the membrane-associated ERα in viscerally-specific neurons. PGE2 could potentiate the response of the TRPV1 and E2 effects on TRPV1 under pathological conditions such as inflammation might be one of the pain signaling pathways. This data confirmed that the role of ERα and not ERβ in the E2 modulates PGE2-mediated TRPV1 activation in visceral DRG neurons. These present situation can reflect patients suffering from pathological abdominal pain in clinical studies for many functional disorders.

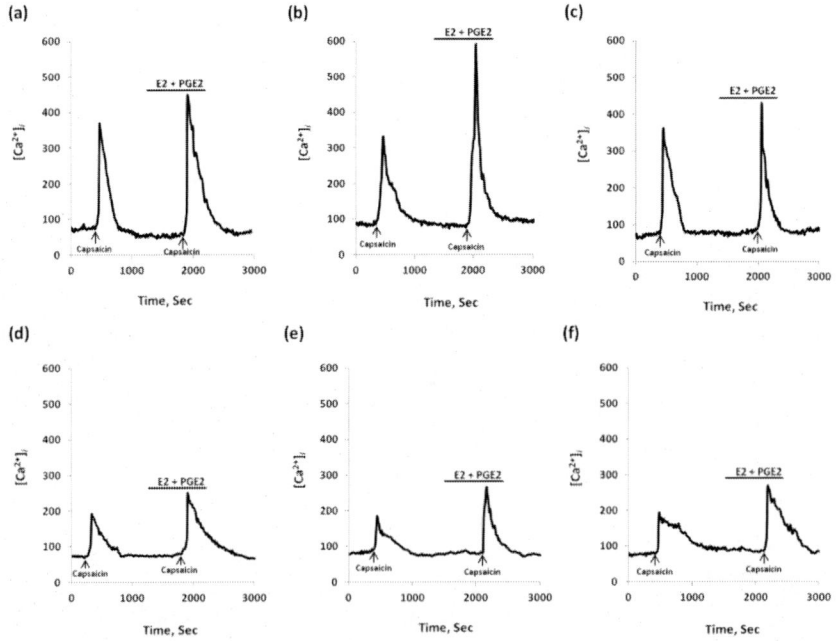

Figure 6.5. The effect of 17β-estradiol (E2) on PGE2 potentiation of capsaicin-induced $[Ca^{2+}]_i$ fluxes in visceral and somatic DRG neurons from wild type, ERαKO, and ERβKO mice. E2 pretreatment with PGE2 significantly decreased capsaicin-induced $[Ca^{2+}]_i$ fluxes in visceral DRG neurons from (a) wild type and (c) ERβKO mice. (b) In visceral DRG neurons from ERαKO mice, E2 didn't inhibit PGE2-potentiated capsaicin-induced $[Ca^{2+}]_i$ fluxes. In somatic DRG neurons, there are no significant attenuation and low amplitudes on capsaicin-induced $[Ca^{2+}]_i$ in (d) wild type, (e) ERαKO and (f) ERβKO mice compared with visceral DRG neurons.

The mechanisms of viscero-visceral cross-sensitization play an important role in the demonstration of visceral pain syndromes affecting different organs often coexist. Although it is usually understood that each primary sensory neuron is a single channel, several studies have demonstrated that DRG neurons can innervate both visceral and somatic organs. This study using retrograde labeling from uterus and colon gave important challenges that the same sensory neuron can innervate both reproductive and gastrointestinal organs. These dichotomized fibers provide a novel mechanism for sensitization of one internal organ by another. Our study show that DRG neurons that innervate both colon and uterus in short-term culture express capsaicin-sensitive TRPV1 receptors and ATP-sensitive P2X3/α,β-meATP-sensitive P2X2/3 receptors which mediate the response to putative nociceptive signals. Our findings suggest that sensory information to the sensory neurons may initiate in different viscera. ER agonists mimic activation of these receptors *in vivo*. Moreover, these neurons can be a good model of endogenous signals. Many literatures suggest that E2 modulation of nociceptive pain processing in functional pain disorders. Within our hypothesis, E2 modulation depends on pain types, durations, and other anti-nociceptive signaling pathways. Thus, E2 modulates short-cultured sensory DRG neurons to capsaicin and ATP/α,β-meATP, suggesting that visceral nociceptive neurons are modulated by E2, which can manifest clinical findings, animal sex-differences associated with visceral hyperalgesia, and a potential target for mediating nociception. In addition, E2 modulation of viscero-visceral cross sensitization appear in DRG neurons [4-5] and P2X3/P2X2/3 and TRPV1 receptors in retrogradely- labeled visceral DRG neurons attenuated ATP/α,β-meATP/capsaicin-induced $[Ca^{2+}]_i$ fluxes, but retrogradely- labeled somatic DRG neurons didn't inhibit $[Ca^{2+}]_i$ responses, suggesting that nociceptive signals modulate peripherally visceral pain sensitization *in vitro*.

Nociceptive mechanisms including the functional pain syndromes are complicated by co-morbid disorders. Visceral nociception and nociceptor sensitization appear to be regulated by ATP/capsaicin. The medium (diameter < 40 microns) and small (diameter < 25 microns) diameter DRG neurons in short-term culture respond to ATP/capsaicin and were estrogen-sensitive, indicating that there may be cross-activation of these receptors that may underlie the pain perception of visceral nociceptors. Although large numbers of clinical and animal studies have indicated that there are sex and estrous cycle differences in pain perception of various pain models, it remains unclear whether sex and estrous cycle differences in ATP/capsaicin-induced acute pain also occur in mice. Furthermore, the PGE2 is synthesized and released in response to damaged tissues, contributes to hyperalgesia, and is involved in the acute and chronic inflammatory reactions. The E2 is anti-nociceptive and decrease capsaicin-induced $[Ca^{2+}]_i$ responses without PGE2 present. In our previous studies showed that PGE2 enhanced the Ca^{2+} responses induced by ATP and capsaicin. This data indicated that the sensitizing actions of PGE2 on retrogradely labeled visceral DRG neurons are mediated through the pain signaling pathway such as cAMP/PKA. Also, recent studies show that E2 acts through an ERα in modulating the TRPV1 receptor-mediated $[Ca^{2+}]_i$ response in retrogradely- labeled visceral DRG neurons, since its effect was eliminated in ERαKO mouse and retained in ERβKO. The present results demonstrate an important non-reproductive role of ERα in modulating capsaicin-induced Ca^{2+} signaling at the level of the visceral afferent neuron, thereby modulating the sensitivity to painful stimuli in the periphery. The abdominal pain related with irritable bowel syndromes and acute and chronic/repeated pelvic pain in women are good examples of visceral pain associated with sensitization [6, 7].

Patients who has pain related with irritable bowel syndrome show the most depressing and see a physician for consulting about this major factor. Hence, from a public health view, this results will have a important impact and improve our knowledge of nociceptive functional disorders including irritable bowel syndrome, chronic pelvic pain, and interstitial cystitis. Moreover, these will help accomplish a deeper understanding of sex-related differences in clinical studies of these diseases in various psychiatric diseases. Only these understanding of implicated mechanism can designs of new and more efficient clinical therapies. Women and men were response to many illnesses differently. It is well categorized that differences between sexes and diseases, disorders, and conditions.

REFERENCES

[1] Lee O., Mayer E., Schmulson M, Chang L., Naliboff B. Gender-related differences in IBS symptoms. *American Journal of Gastroenterology,* 2001, 96, 7: 2184- 2193.

[2] Dray A. Neuropathic pain: emerging treatments. *British Journal of Anesthesia,* 2008, 101, 1: 48-58.

[3] Gold M., Dastmalchi S., Levine J. Co-expressed of nociceptor properties in dorsal root ganglion neurons from the adult rat in vitro. *Neuroscience,* 1996. 71, 1: 265-275.

[4] Chaban, V., and Micevych P. Estrogen receptor -alpha mediates estradiol attenuation of ATP-induced Ca^{2+} signaling in mouse dorsal root ganglion neurons. *Journal of Neuroscience Research* 2005. 81, 1: 31-37.

[5] Sarajari S., Oblinger M. Estrogen effects on pain sensitivity and neuropeptide expression in rat sensory neurons. *Experimental Neurology* 2010. 224(1): 163-169.

[6] Giamberardino M. A. Recent and forgotten aspects of visceral pain. *European Journal of Pain* 1999. 3, 2: 77-92.

[7] Chaban V. Editor (2017) *Irritable Bowel Syndrome: Novel Concepts for Research and Treatment*, InTech, 94p.

Chapter 7

CALCIUM SIGNALING IN ASTROCYTES

ABSTRACT

Appreciating the physiology of astrocytes and their role in brain functions requires an understanding of molecules that activate these cells. Estradiol may influence astrocyte functions. We now report that estrogen altered intracellular calcium concentration ([Ca2+]i) in neonatal astrocytes that ex- pressed estrogen receptor (ER) mRNA in vitro. Western blotting revealed both ERα and ERβ proteins in both the nuclear fractions and plasma-membrane fractions.

Application of 17β-estradiol (20 nM) to fura 2-loaded astrocytes in vitro stimulated [Ca2+]i in 75% of astrocytes with an EC50 of 12.7 + 3.1 nM.

This rapid action of estradiol was blocked by the ER antagonist, ICI 182,780. Removal of extracellular Ca2+ did not alter the effect of estradiol, but phospholipase C inhibitor U73122 (10 μM) and 2-aminoethoxydiphenyl borate (5 μM), an inhibitor of the inositol-1,4,5,-trisphosphate-gated intracellular Ca2 □ channel, significantly decreased the estradiol-induced [Ca2+]i flux.

Estradiol was unable to induce [Ca2+]i flux in thapsigargin- depleted cells. These results indicate that estradiol mediates [Ca2+]i flux in astrocytes through a membrane-associated ER that activates the phospholipase C pathway.

EXPRESSION OF ERα AND ERβ IN NEONATAL ASTROCYTES

Astrocytes appear to express estrogen receptors (ER) *in vivo* and *in vitro* [1] and may mediate a number of estradiol-induced effects in the brain. For example, ERs acting through astrocytes have been implicated in estrogen action on synaptic plasticity and neural repair [2]. Moreover, astrocytes may have a central role in sexual differentiation of the brain and in reproduction through modulation of the estrogen-positive feedback in the luteinizing hormone (LH) surge [3].

In addition to the transcriptional effects of estradiol via nuclear ERs, estradiol may rapidly activate cells by increasing cytoplasmic $[Ca^{2+}]_i$ levels. In addition to the transcriptional effects of estradiol via nuclear ERs, estradiol may rapidly activate cells by increas- ing cytoplasmic $[Ca^{2+}]_i$ levels.

Primary Cell Culture

Enriched astrocyte cultures were derived from the Long-Evans rat neonatal cortex (1–2 d old) using a technique modified from McCarthy and de Vellis (28) that favors the survival and proliferation of astrocytes over neurons. Briefly, glial cells dissociated with papain were grown in DMEM/F12 supplemented with 10% fetal calf serum (Globepharm, Deerfield, IL).

Oligodendrocyte precursors (top cells) and microglia were separated from the glia by shaking after 24 h *in vitro*. This process was repeated until the purity of the cell culture was more than 95% astrocytes, assessed immunocytochemically with the astrocyte markers galactocerebroside and glial fibrillary acidic protein (GFAP). The remaining 5% of non-GFAP-positive cells could include astrocytes, oligodendrocytes, and microglia but not neurons because neuronal-specific enolase immunocytochem- istry did not label any cells in the cultures. For Ca^{2+} imaging studies, the cultures were grown to confluence and subcultured using an EDTA- trypsin wash. DMEM medium was removed from flasks, and cells were washed twice with 5 ml EDTA-trypsin, 2.5% trypsin solution, and then resuspended in 8 ml DMEM/F12–10% fetal calf serum. Astrocytes were centrifuged for 5 min at 80 X g and plated onto Matrigel (Beckton Dickinson, Franklin Lakes, NJ)-coated, 15-mm coverslips (Life Technologies, Inc., Grand Island, NY). Before experiments, all cells were gently washed with serum- and phenol red-free DMEM/F12 (free DMEM) and then incubated overnight in the same media.

RT-PCR Analysis

Total RNA was isolated from several cultures of Long-Evans rat neonatal astrocytes using Trizol Reagent (Invitrogen Corp., Carlsbad, CA) according to the manufacturer's protocol. RNA was further purified using the Absolutely RNA RT-PCR Miniprep Kit (Stratagene, La Jolla, CA) with an addition of a deoxyribonuclease I digestion step to remove genomic DNA. RNA integrity was confirmed by 2% agarose gel elec- trophoresis. cDNA was synthesized from the total RNA using Super- Script III First-Strand Synthesis System for RT-PCR (Invitrogen) and subjected to PCR using specific primers for ERα and ER genes. Primers were designed using Primer3 software (The Whitehead Institute, Boston, MA), and their specificity was confirmed by a BLAST software-assisted search of a non-redundant nucleotide sequence database (National Library of Medicine, Bethesda, MD). PCR experiments were conducted on iCycler iQ Real-Time PCR Detection System (Bio-Rad Laboratories, Hercules, CA) using Brilliant SYBR Green QPCR Master Mix (Stratagene). Each cycle consisted of the following three steps: denaturation for 45 sec at 95 C, annealing for 1 min at 56 C, and 1 min of elongation at 72 C. Data were collected in real time during the elongation step after each cycle of the 40-cycle reaction. The dissociation/melting curve analysis performed after the reaction was completed indicated presence of a single product with a melting temperature of 83 C in the samples processed with reverse transcriptase. Negative controls without reverse transcriptase and water controls were included in each reaction. In addition to real-time and melting curve analysis of the reactions, amplified products were separated electrophoretically in 3% agarose gels with ethidium bromide and visualized and photographed under UV light to confirm the proper size and the absence of nonspecific products. As an additional control, sam- ples were purified using Rapid PCR Purification System (Marligen Biosciences, Ijamsville, MD) and sequenced at the University of California, Los Angeles, DNA Sequencing Core Facility.

Western Blotting

Western blots were probed with rabbit monoclonal primary antibody to ERα (clone SP1, Lot no. 9101-50-26; Lab Vision Corp., Fremont, CA) and rabbit polyclonal antibody (Lot no. 24916; Upstate Biotech- nology, Inc, Charlottesville, VA). Membrane proteins were separated from the intracellular proteins using Mem-PER(r) Eukaryotic Membrane Protein Extraction Reagent Kit (Pierce Biotechnology, Rockford, IL). Thirty microliters of each fraction were used for immunoblot analysis. The detection was done using a donkey antirabbit IgG secondary antibody (Jackson ImmunoResearch, West Grove, PA) and developed using SuperSignal West Pico Chemiluminescent Substrate (Pierce Biotechnology). Both ERα and ERβ staining was blocked by preincubating each antibody with its immunizing peptide as described in the enclosed literature. Astrocyte cultures were washed three times with PBS, and then cells were homogenized at 2000 rpm in 50 mm Tris-HCl (pH 7.5) containing general-purpose protease inhibitor cocktail (Sigma Chemical Co., St. Louis, MO.) Nuclear pellets were collected through low-speed centrifugation. The remaining supernatant was centrifuged at 49,000 μ g for 15 min at 4 C. The supernatant, containing nonmembrane proteins, was collected. The pellet was washed with 300 l of 50 mm Tris containing protease inhibitors and centrifuged at 49,000 μ g for 15 min at 4 C to pellet the membranes. Eighty micrograms of total protein from each fraction were blotted onto Invitrolon PVDF Membrane (Invitrogen) and probed with antibodies against ERα and ERβ.

Digital Fluorescence Videomicroscopy

$[Ca^{2+}]_i$ was measured by the ratiometric method. Coverslips were mounted in an Attofluor recording chamber, and changes in $[Ca^{2+}]_i$ were measured by the Attofluor Ratio Vision Digital Fluorescence Microscopy System (Atto Instruments, Rockville, MD). After incubation with the fluorescent marker (fura 2, 5 μm; Molecular Probes, Eugene, OR) for 45 min at 37 C, the cells were washed and kept in Hanks' balanced salt solution (HBSS; Life Technologies) for the length of the experiment. Astrocytes were bathed and perfused (1–5 ml/min) with HBSS buffered with HEPES (20 mm) using a peristaltic pump (Rainin Instrument, Woburn, MA). Water- soluble 17β-estradiol (cyclodextrin-encapsulated 17□-estradiol; Sigma), thapsigargin (Sigma), and 1,3,5(10)-estratrien-3,17-diol-6-one:BSA (E- 6-BSA; Steraloids Inc., Newport, RI) were applied by brief superfusion (10 sec) of the experimental chamber with HBSS until an increase in $[Ca^{2+}]_i$ was detected. A $[Ca2□]i$ increase that exceeded 50 nm was considered a $[Ca^{2+}]_i$ transient. ICI 182,780 (Tocris Cookson, Ellisville, MO) and 2-aminoethoxydiphenyl borate (2-APB; Sigma) were added 5 min before the addition of estradiol. In the experiments testing the contribution of extracellular Ca^{2+}, cells were incubated in a Ca^{2+}-free HBSS medium with 10 mm of 1,2-bis(2-aminophenoxy)ethane-N-tetra-acetic acid (Sigma) replaced the 1.8 mm $CaCl_2$. To demonstrate that a membrane-associated ER induces $[Ca^{2+}]_i$ flux, E-6-BSA, filtered through a 3-kDa cut-off filter (Amicon, Beverly, MA) to remove free 17β-estradiol, was immediately applied to cultured astrocytes by rapid perfusion. All experiments were performed at room temperature (20 –23° C).

RT-PCR analysis of astrocytic mRNA demonstrated the expression of ERα and ERβ. Agarose gel electrophoresis of the PCR products indicated the presence of the appropriately sized bands in samples with reverse transcriptase and the absence of amplicons in the samples processed without re- verse transcriptase and in water controls (Figure 7.1). Samples were normalized to 18S rRNA. Real-time RT-PCR analysis determined the threshold cycle to be between cycles 30 and 32 for both ERα and ERβ. RNA samples of ovarian tissue, used for positive control, had a threshold cycle between cycles 25 and 26. Compared with the threshold cycle for ovary, this result suggests that astrocytes have a low level of ERα and ERβ mRNA. Combined GFAP immunocytochem- istry and *in situ* hybridization for ERα and ERβ demonstrates that both ERs are expressed in GFAP-positive astrocytes.

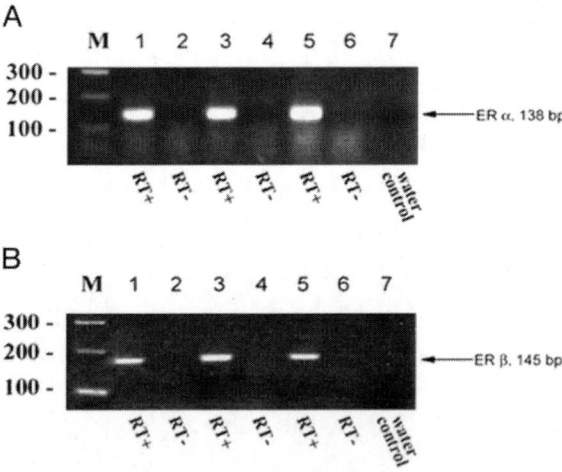

Figure 7.1. RT-PCR analysis of ERα and ERβ gene expression in rat neonatal astrocytes. Samples with reverse transcriptase show a band of 138 bp corresponding to the amplicon of ERα (A) and a band of 145 bp corresponding to ERβ (B), which are absent from the same samples processed without the reverse transcriptase and in the water controls. Lane M, 100 bp DNA marker; lanes 1 and 2, sample 1; lanes 3 and 4, sample 2; lanes 5 and 6, sample 3; and lane 7, water control [5].

EFFECTS OF ESTRADIOL ON $[CA^{2+}]_I$ IN NEONATAL ASTROCYTES *IN VITRO*

Approximately 75% of the astrocytes responded to 20 nm 17β-estradiol stimulation within 10 –30 sec of treatment. The initial $[Ca^{2+}]_i$

transient peaked at 252.9 ± 18.3 nm (mean ± sem; n = 53). Although there was variability in the shape of the falling phase, typical $[Ca^{2+}]_i$ transients rapidly reached a maximum and then slowly decayed to baseline levels (t1/2 21.5 ± 3.6 sec; Fig 7.2). The estradiol-induced $[Ca^{2+}]_i$ levels in astrocytes were dose responsive (1 nm, 20 nm, 50 nm, and 100nm) and a maximum response at 50 nm estradiol. Based on preliminary studies, a 5- to 10-min period between stimuli was sufficient to allow for refilling of the $[Ca^{2+}]_i$ stores and to induce a similar $[Ca^{2+}]_i$ increase with the estradiol. After a 5- to 10- min wash-out period, the second estradiol-induced $[Ca^{2+}]_i$ transient was statistically the same as the first ($P < 0.05$; Figure 7.2 A). A comparison between the first and second stimuli was used to measure and compare the amplitude of the drug effects as previously described [4].

Estradiol-induced $[Ca^{2+}]_i$ transients (Figure 7.2) were heterogeneous in terms of amplitude, induction, and frequency of oscillations, an observation that is consistent for $[Ca^{2+}]_i$ flux induced by other stimuli such as ATP. In 25% of the estradiol-responsive astrocytes, a single pulse of estradiol induced a series of $[Ca^{2+}]_i$ transients com- posed of two or more $[Ca^{2+}]_i$ spikes with an average duration of 3–5 min (Figure 7.2 B). The $[Ca^{2+}]_i$ spikes in these oscillations had a periodicity of 57.05 ± 3.9 sec. Of the astrocytes in which estradiol induced oscillations, 70% had successively smaller transients (Figure 7.2 B), and 30% had oscillatory transients that had spikes all of similar amplitude.

ESTRADIOL RELEASES CA^{2+} FROM INTRACELLULAR STORES

To determine whether estradiol-induced $[Ca^{2+}]_i$ transients resulted from Ca^{2+} influx of extracellular sources or from mobilizing intracellular $[Ca^{2+}]_i$ stores, the response to estradiol was tested in astrocytes perfused with HBSS containing 10 mm of 1,2-bis(2-aminophenoxy)ethane-N,N,N,N-tetraacetic acid but no Ca^{2+}. The estradiol-induced $[Ca^{2+}]_i$ spikes were not significantly different in Ca^{2+}-free media from those in Ca^{2+}-containing media (224.5 ± 22.7 nm, n = 12, Figure 7.2 A *vs.* 252.9 ± 18.3 nm, n = 53, $P < 0.05$, suggesting that estradiol predominantly mobilized Ca^{2+} from intracellular stores. Further support for this interpretation was obtained from astrocytes in which thapsigargin (m) depleted intracellular $[Ca^{2+}]_i$ stores. After intracellular $[Ca^{2+}]_i$ depletion, estradiol did not elicit a $[Ca^{2+}]_i$ spike (Figure 7.2 D). These results support the hypothesis that the main source of the initial estradiol-induced $[Ca^{2+}]_i$ transient was intra- cellular.

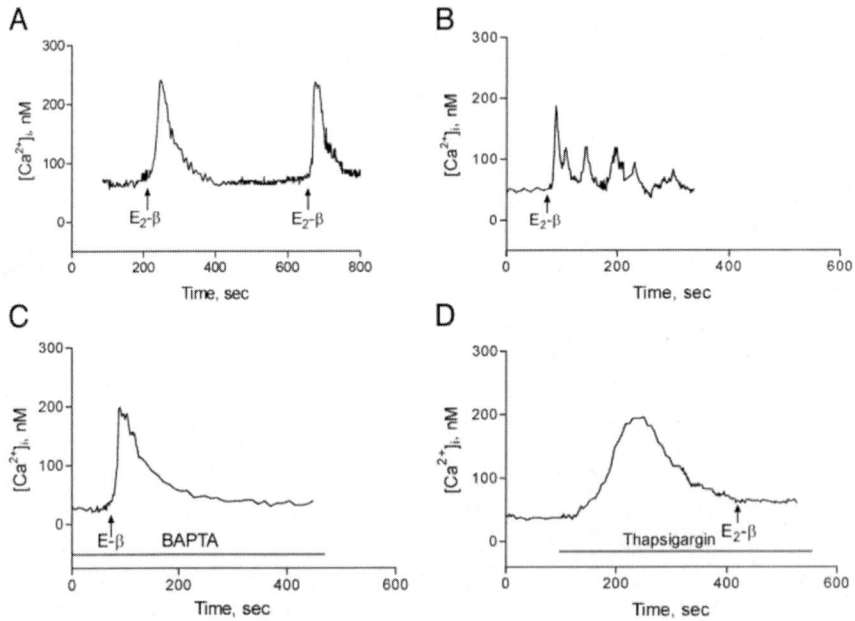

Figure 7.2. Effect of 17β-estradiol (E2) on $[Ca^{2+}]_i$ in astrocytes. A, Representative $[Ca^{2+}]_i$ levels in cortical astrocytes stimulated with E2 (20 nM); the *arrows* in each trace indicate time of addition. The effect of E2 (20 nM) was fully reversible after 10 min of washout. B, In a subpopulation of astrocytes (25%), E2 stimulated Ca^{2+} spikes. C, Representative trace of $[Ca^{2+}]_i$ in astrocytes stimulated with E2 in Ca2☐-free medium (in the presence of 10 mM 1,2-bis(2-aminophenoxy)ethane-*N,N,N,N*-tetraacetic acid) indicates Ca^{2+} release from intracellular stores. D, Endoplasmic reticulum Ca^{2+}-ATPase inhibitor, thapsigargin (1 M), induced $[Ca^{2+}]_i$ increase and abolished subsequent E2 (20 nM) stimulation [5].

Although estradiol can elicit a release of Ca^{2+} through the formation of IP3 in neurons, and astrocytes (present study), the predominant effect of estradiol in neurons has been to inhibit L-type voltage-gated Ca^{2+} channels.

This has been demonstrated in neostriatal, hippocampal, and dorsal root ganglia neurons. In neurons, estradiol did not activate $[Ca^{2+}]_i$ directly, but rather, it attenuated the induced $[Ca^{2+}]_i$ flux. The pharmacology of the astrocytic ER and the neuronal ER that mediate rapid $[Ca^{2+}]_i$ transients were similar; the receptors were stereospecific and blockable by ICI 182,780. In neurons, however, the $[Ca^{2+}]_i$ flux is dependent on extracellular Ca^{2+}, but in astrocytes, estradiol stimulated the release of $[Ca^{2+}]_i$. Several possible explanations can account for these differences. There is an evidence for both membrane-associated ERα and E; however, it is possible that in astrocytes ERα may be the predominant ER expressed in the membrane. Thus, rapid action of estradiol may be mediated by different ERs. Another explanation is that ERs may be coupled to different signaling cascades in astrocytes *vs.* neurons. Currently, neither hypothesis can be formally excluded. Regardless of which ER mediated these actions, the present results demonstrate direct, rapid, and reversible estradiol-mediated $[Ca^{2+}]_i$ signaling in astrocytes, supporting the conclusion that there is a mechanism through which estradiol can modulate astrocyte excitability. In addition, ATP seems to be the most abundant and reliable $[Ca^{2+}]_i$ mobilizer in macroglial cells.

REFERENCES

[1] Sato K, Matsuki N., Ohno Y., Nakazawa K. Estrogens inhibit l-glutamate uptake activity of astrocytes via membrane estrogen receptor. *Journal of Neurochemistry,* 2004, 86: 1498 – 1505.

[2] Dhandapani K. and Brann D. Estrogen-astrocyte interactions: implications for neuroprotection. *BMC Neuroscience* 2002, 3: 6.

[3] Micevych P, Sinchak K, Mills R, Tao L, LaPolt P, Lu J. The luteinizing hormone surge is preceded by an estrogen-induced increase of hypothalamic progesterone in ovariectomized and adrenalectomized rats. *Neuroendocrinology,* 2003, 78: 29 –35.

[4] Chaban V., Mayer E., Ennes H., Micevych P. Estradiol inhibits ATP-induced intracellular calcium concentration increase in dorsal root ganglia neurons. *Neuroscience,* 2003, 118: 941–948.

[5] Chaban V., Lekhter A. and Micevych P. A membrane estrogen receptor mediates intracellular calcium release in astrocytes. *Endocrinology,* 2004, 145(8): 3788–3795.

Chapter 8

CALCIUM SIGNALING IN VISCERAL NOCICEPTION

ABSTRACT

Dorsal Root Ganglia (DRG), containing the soma of sensory afferent neurons, function as a possible new target for neuromodulatory devices to treat visceral pain. The modulating effect of DRG activation through changes in $[Ca^{2+}]_i$ could be validated by using a behavioral assay of viscero-motor responses *in vivo*. We focused on etiology of Irritable Bowel Syndrome (IBS), Painful Bladder Syndrome (PBS), Chronic Pelvic Pain Syndrome (CPP), fibromyalgia and other as examples of functional disorders characterized by clinical presentation of visceral pain.

The incidents of persistent visceral pain associated with IBS, CPP, PBS and others is 2- 3 times higher in women than in men, suggesting estrogen modulation. Our previous studies strongly suggest that this modulation occurs through changes in Ca^{2+} signaling.

BASIC RESEARCH CONCEPTS AND THERAPEUTIC INTERVENTIONS IN VISCERAL PAIN

The accumulation of nociceptive diseases that limit normal body functions is a major risk factor for a disability, and visceral pain is one of the most prevalent human health problems. Additionally, many pain-associated diseases are accompanied by the concomitant decline in cognitive and motor performance. The complex interplay and balance between diverse signal transduction mediators, genetic background and environmental factors may ultimately determine the outcome of nociceptive progression in various disorders. Pain is a subjective feeling that is difficult to standardize and parameterize in a traditional fashion for scientific analysis. The causes of visceral pain are often not very clear, as there are many symptoms of the reproductive, gastrointestinal, musculoskeletal, neurological, psychological systems and urinary tract that often co- occur in the same patient. VIscero-somatic and viscero- viseral hyperalgesia and allodynia Cause the perception of pain to spread for an initial area, to adjacent visceral sites [1].

Often times, there is not a clear relationship between the severities. There is often no clear relationship between the severity of the visceral pain and pathology in the viscera, including the reproductive tract, urinary bladder and colon. The clinician treating this pain is often times tempted to adopt a unidimensional approach, focusing on one organ system, ignoring the psychological and behavioral manifestations of the visceral pain. Therefore, studies of the nervous system in individuals with visceral pain associated with reproduction such as CPP, urinary system such as PBS and bowel disorders such as IBS, suggest a model in which alteration in the central stress circuits in predisposed individuals may trigger and then maintain, the pain and pathophysiological changes in the viscera [2]. These patients have significantly more depression, psychological and somatic complains and more often give a history of physical, sexual or emotional abuse, or trauma. Chronic visceral pain results in adverse affects not only one's mood, but also their professional and social lives, as well as general well- being. Pain accounts for a majority of all primary health care visits. For the past decade, medical literature has carefully documented the under-treatment of all types of pain by physicians. Pain is a complex and individual experience that is often difficult for patients to fully describe using a conventional clinical assessment [3]. Visceral pain affects up to 25% of women at some time in their lives (about a billion worldwide) and can result in dysmenorrhea, dyspareunia, menstrual irregularities, back pain, gastrointestinal and genitourinary symptoms and reduced fecundity.

There are two essential components of pain: discriminative and affective. The discriminative component includes the ability to identify the stimulus as originating from somatic or visceral tissue, determine some of the physical properties of the stimulus and localize it in space, time and along a continuum of intensities. The affective component is the experience, which motivates escape, avoidance and protective behavior. All of these components of pain must be considered in any discussion of the neurophysiological basis of visceral pain. Because of the inherent subjectivity of pain, there is a wide disparity among individuals in the way that they experience pain generated by what seem to be similar stimuli. There is also a tension between the subjectivity of the patient's pain experience and the common insistence of the clinician upon objective findings that are proportionate with the patient's complaints, to enable to distinguish between exaggerated pain reports. Proposed therapeutic considerations must also include the neural systems modulating pain, for it is well known that pain can be profoundly influenced by other somatic stimuli and by attentional, emotional and cognitive factors.

An important focus of clinical management now includes the assessment of pain on various aspects of a patient's existence. The health-related quality of life that encompasses Health related qualities of life are comprised of aspects of health and well-being that are valued by patients, such as their emotional, physical, and cognitive state, and ability to participate in meaningful tasks. There is a concern that not enough emphasis is placed on a clinical validity (i.e., issues which are important to patients and reflect their experiences). A balance between biomedical, organ-oriented and cognitive interpersonal approaches is the most appropriate to study this psychosomatic interface. In view of the iatrogenic component in the maintenance of painful syndromes, clinician-centered interventions and close observation of the clinician-patient relationship are of particular importance. Nociceptive responses involve a vast number of messenger molecules that interact with enzymes and receptors of all classes. They direct the recruitment of different types of cells to assist in the recovery of a health state. A balance between these messengers and the redundancy of various body systems presents major difficulties for therapeutic intervention. Nevertheless, it is a very important aspect to consider in the treatment of disorders association with visceral pain.

THE INFLUENCE OF GONADAL HORMONES ON CA^{2+} SIGNALING ASSOCIATED WITH VISCERAL PAIN PERCEPTION

Most functional syndromes are associated with pain, which is the symptom that patients list as the most depressing and it is a major factor for consulting a physician. Estrogen receptor alpha (ERα) plays a significant role in modulating pain signaling. In females, 17β-estradiol (E2) that activates ERα receptor is involved in inflammation and pain. The pain originating from the pelvic structures often overlaps, manifesting a diffuse peritoneal pain. Typically, this type of pain is functional in nature without any clear pathological manifestation in the organ. However, the pain due to dysfunction of a specific pelvic organ, such as inflammation, obstruction or stricture, can overlap with other organs. Commonly, the overlapping of pelvic pain occurs between the urinary bladder, lower gut and uterine inflammation.

Sex is a biological variable that is frequently ignored in study designs and analyses. Estrogen has a significant role in modulating visceral sensitivity, indicating that sex steroid alterations in sensory processing may underlie sex-based differences in functional pain symptoms. However, reports of estrogen modulation of visceral and somatic nociceptive sensitivity are inconsistent. To help resolve these inconsistencies, the clinical and scientific community needs to focus on sex steroid actions on nervous system. Little is known about estrogen-mediated mechanisms in peripheral nervous system, but the fact that DRG neurons express both estrogen receptors (ERα and ERβ) and respond to estrogen treatment by modulating different nociceptive pathways suggest a potential target for pain treatment.

In our studies we observed that DRG neurons innervating viscera have a greater $[Ca^{2+}]_i$ response to subsequent ATP and capsaicin and NMDA stimulation than somatic afferents [4]. These observations indicate that viscerally- specific neurons express receptors with higher permeability to Ca^{2+}, which can modulate transduction of nociceptive signals and suggest that *visceral* afferents are functionally different from *somatic* afferents.

Sensitization of primary afferent neurons may play a role in the enhanced perception of visceral sensation leading to pain. Endometriosis, acute and recurrent/chronic pelvic pain in women or abdominal pain from IBS are all visceral pain sensations that may result in part from sensitization.

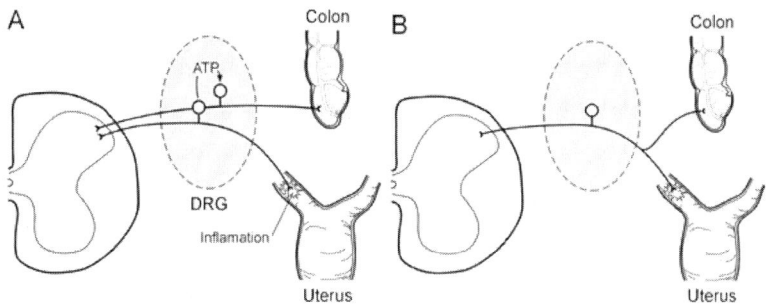

Figure 8.1. Non exclusive models of two possibilities for viscero-visceral cross-sensitization in the DRG. In (A), ATP released by a neuron innervating the inflamed uterus acts on a neighboring neuron sensitizing its responses to colonic distention. In (B), the same neuron innervates the uterus and colon. Uterus inflammation directly sensitize the neuron to colonic distention [4].

17β – estradiol (E2) acts in DRG neurons to modulate L-type voltage-gated calcium channels (VGCC) and through group II metabotropic glutamate receptors. E2 has a significant role in modulating visceral sensitivity, indicating that E2 alterations in sensory processing may underlie sex- based differences in functional pain symptoms. However, reports of E2 modulation of Ca^{2+} signaling on visceral and somatic nociceptive sensitivity are inconsistent. Dorsal root ganglia (DRG), the clustering of sensory neural cell bodies) have emerged as a promising neuromodulatory target to manage certain types of visceral pain according to clinical evidence. But it remains unknown whether and in what mechanisms visceral pain could be effectively attenuated by DRG stimulation. Our knowledge of visceral sensory neurons relevant to DRG neuromodulation of $[Ca^{2+}]_i$ is scarce, which prevents the further development of the technique to benefit patients with chronic visceral pain.

Compared with non-visceral ('somatic') pain, visceral pain is difficult to localize and is referred to other tissues. Sensory input from healthy viscera gives rise to minimal conscious sensation, and stimuli (e.g., burning, pinching) that evoke pain from the skin fail to be noticeably perceived from the viscera. When viscera are diseased or inflamed, however, stimuli that normally produce innocuous sensations can become painful. Like many chronic pain conditions, prolonged pain associated with functional disorders is initiated by activity in peripheral sensory (afferent) neurons. Afferent sensitization (i.e., increase in excitability) is manifested by a reduced response threshold, enhanced response magnitude, and/or recruitment of mechanically- insensitive afferents to mechanosensitivity [5]. Both low- and high-threshold colorectal afferents sensitize or acquire mechanosensitivity after brief exposure to inflammatory mediators. Visceral afferent sensitization appears necessary and sufficient for chronic pain in etiology of functional disorders and pharmacological treatment of visceral pain is unsatisfactory. Accordingly, developing non-pharmacological alternatives for managing visceral pain states is evolving. Ca^{2+} studies may reveal complex response profiles of primary afferents to different nociceptive stimuli.

Similar to other chronic diseases, a multicomponent conceptual model of functional disorders involves physiologic, cognitive and behavior factors will be necessary for developing new therapies. The different systems such as neuroendocrine regulation and autonomic response will affect nociceptive systems. From a public health perspective, a substantial impact on our knowledge of nociceptive functional diseases will help achieve a deeper understanding of data presented in clinical aspects of these symptoms. Only a thorough understanding of the mechanism implicated in these phenomena can truly contribute to the designing of new and more efficient therapies.

VISCERO-VISCERAL CROSS-SENSITIZATION

Within the context of the cross-sensitization hypothesis, inflammation sensitizes non-inflamed viscera that are innervated by the same DRG, and/or cross-sensitization occurs as a result of intra-DRG release of sensitizing mediators such as ATP within the DRG (Figure 8.1). According to this new hypothesis, the reproductive tract inflammation sensitizes DRG neurons innervating the visceral organ. Several lines of evidence indicate that E2 directly influences the functions of primary afferent neurons. Both subtypes of estrogen receptors (ERα and ERβ) are present in DRG neurons, including the small-diameter putative nociceptors. *In vitro,* ATP-sensitive and vanilloid-sensitive DRG neurons respond to E2, which supports the hypothesis that visceral afferents are E2 sensitive: i) visceral pain is affected by hormonal level in cycling females; ii) there are sex differences in the prevalence of functional disorders involving the viscera and iii) putative visceral afferents fit into the population of DRG neurons that are sensitive to E2. These data suggest that in addition to CNS actions, E2 can act in the periphery to modulate nociception.

Chronic pain management is a major scientific and public health care challenge, as current analgesic drugs rarely provide sufficient efficacy in the absence of serious side effects. Sensitivity to pain remains long after tissue healing. The discovery of the neurochemical mechanisms that maintain chronic pain hypersensitivity is needed for a better treatment. The outcome of the studies focused on new mechanisms of visceral pain will have a substantial impact on our knowledge of pain-associated diseases and may help to achieve a deeper understanding of gender differences presented in clinical aspects of the functional syndromes such as irritable bowel syndrome, chronic pelvic pain, interstitial cystitis/painful bladder syndrome, fibromyalgia and others. Chronic visceral pain represents a significant health problem worldwide with women being disproportionately affected, and constitutes a serious threat associated with overuse of anti-pain medications. Treatment options for chronic pain are often limited and significant side effects include risk of addiction. Therefore, new pain therapies based on detailed understanding of chronic pain mechanisms are needed as alternatives to current analgesics.

REFERENCES

[1] Pan, X. and Malykhina A., Estrous cycle dependent fluctuations of regulatory neuropeptides in the lower urinary tract of female rats upon colon-bladder cross-sensitization. *PLOS One,* 2014, 9: 94872-94872.
[2] Mayer, E. Gut feelings: The emerging biology of gut-brain communication. *Nature Review Neuroscience,* 2011, 12: 453- 466.
[3] Meltzak, R. Pain and the Neuromatrix in the Brain. *Journal of Dental Education,* 2001, 65: 1378-82.
[4] Chaban V. Estrogen modulation of visceral nociceptors. *Current Trends in Neurology,* 2014, 7: 51- 56.
[5] Bielefeldt K, and Gebhart G. Visceral Pain - Basic Mechanisms. In: *Wall and Melzack's Textbook of Pain* (Koltzenburg M, and McMahon S.- Editors) Churchill-Livingstone: Saunders, 2013.

Chapter 9

CALCIUM SIGNALING IN CHRONIC AND NEUROPATHIC PAIN

ABSTRACT

Neuropathic and chronic pain symptoms account up to 50% of all visits to seek health care providers, therefore development of appropriate therapies is of great clinical and societal importance. Despite a successful reduction of pain using during the novel available treatments pain returns in up to 75% of treated patients. Pain is strongly associated with other diseases that can lack of awareness to its pathology is further illustrated by the fact that the average time duration between the onset of pain and the diagnosis is couple of years despite the fact that majority of patients with chronic pain suffer. Patients with chronic pain frequently have pain from several visceral organs. The most common diagnoses include endometriosis, Painful Bladder Syndrome (PBS), Irritable Bowel Syndrome (IBS), Inflammatory Bowel Disease (IBD), pelvic floor tension myalgia, vulvar vestibulitis, and vulvodynia and others. Frequently, pain does not correlate with pathologic findings at the time of clinical assessment. In addition, alterations in the limbic and sympathetic nervous system and hypothalamic-pituitary-adrenal axis mediate a cycle of hypervigilance for pain sensations from different organs, which can lead to descending induction of pathologic changes in these organs. Novel data from our and other laboratories strongly suggest that some tropic and physiological changes associated with chronic indeed are mediated through Ca^{2+}.

Nociceptive Pathways in Clinical Presentation of Diseases Associated with Chronic Pain

Patients with chronic pain frequently have multiple diagnoses. Vicerosomatic and viscero-viseral hyperalgesia and allodynia result in the spread of a perception of pain from an initial site to adjacent areas. Chronic pain may initially have only one pain source in the viscera, such as the uterus in dysmenorrhea or endometriosis implants, but a multitude of mechanisms involving the peripheral and central nervous system can lead to the development of painful sensations from other adjacent organs. Often the etiology of visceral pain is not clear, as there are many symptoms of the reproductive system, gastrointestinal and urinary tracts, musculoskeletal, neurological and psychological systems that often co-occur in the same patient. The variation of pain symptoms and pain perception and behavioral responses to pain in these patients is poorly understood. The treating clinician is often tempted to take a unidimensional approach and focus on one organ system and ignore the psychological and behavioral manifestations of the chronic pain.

The mechanism of endometriosis-induced nociceptive signaling is poorly understood and in some cases pain can be exacerbated by co-morbidity with other chronic pelvic pain syndromes such as irritable bowel syndrome, painful bladder syndrome, vulvodynia and fibromyalgia. It has also been shown that ectopic implants develop sensory nerve supply both in women and in animal models of endometriosis. Sensory input arriving from the visceral organ to the spinal cord divergences at the level of primary sensory neurons which further transmit considerable information from periphery to the central nervous system. Visceral pain may be manifestation from a single organ such as uterus or may arise from algogenic conditions affecting more than one organ [1]. This type of pain is important not only because it is difficult to diagnose its clinical conditions but also for its therapeutic implications. It is quite possible to modulate pain from one viscus to another. Treatment of the endometriotic lesions results in the improvement of spontaneous and referred urinary symptoms [2].

Cross-sensitization in the pelvis implies the transmission of noxious stimuli from one organ to another through an adjacent normal structure resulting in functional (rarely organic) changes. Pelvic organ cross-sensitization is considered as one of the factors contributing to chronic pelvic pain [3]. Chronic pelvic pain (CPP) syndrome affects up to 25% of reproductive age women and results in dysmenorrhea, menstrual irregularities, back pain and reduced fecundity. One of the most common causes of CPP is endometriosis. Chronic pain adversely affects mood, social and professional life and general well being. Thus, assessing the impact of the pain on various domains of a patient's existence has become an important focus in the clinical management. Most women with complaints of pelvic pain will undergo laparoscopy to both diagnose and treat these diseases, but laparoscopy is often is unsuccessful due to lack of intraperitoneal pathology or altered pain processing. Pain out of proportion to identifiable pathology is the most immediate and dramatic consequence of disorders associated with CPP and is responsible for a highly negative impact on quality of life and substantial workforce loss. Results of a national survey determined that 15% of women in the United States have experienced CPP and only 10% of these consulted a gynecologist and 75% did not consult a health care provider of any type. Due to the alarming situation and unmet need, the USA and other countries have launched a call for more focused research on improving the diagnosis and treatment of CPP syndrome.

The pathophysiology of visceral hyperalgesia is less well- known than its cutaneous counterpart, and our understanding of visceral hyperalgesia is colored by comparison to cutaneous hyperalgesia, which is believed to arise as a consequence of the sensitization of peripheral nociceptors due to long-lasting changes in the excitability of spinal neurons. Endometriosis is currently defined as a chronic functional syndrome characterized by recurring symptoms of abdominal discomfort or pain. In the context of visceral pain, the TRPV1 receptor is a sensory neuron-specific cation channel, which plays an important role in transporting thermal and inflammatory pain signals. Evidence for TRPV1's role is that mice lacking TRP1 receptor gene have deficits in thermal- or inflammatory-induced hyperalgesia. Activation of both TRPV1 and P2X receptors induce mobilization of $[Ca^{2+}]_i$ in cultured DRG neurons [4].

Various inflammatory mediators such as prostaglandin E2 (PGE2) and bradykinin potentiate TRPV1. The potentiation of TRPV1 activity can be quantified by measuring the differences of capsaicin-induced Ca^{2+} concentration changes before and after receptor activation [5]. Significantly, a subset of DRG neurons respond to both capsaicin and ATP indicating that there may be cross-activation of these receptors that may underlie the sensitization of visceral nociceptors. Capsaicin-induced TPRV1 receptor-mediated changes in $[Ca^{2+}]_i$ may represent a level of DRG activation to noxious cutaneous stimulation while ATP-induced changes in $[Ca^{2+}]_i$ may reflect the level of DRG neuron sensitization to noxious visceral stimuli since ATP is released by noxious stimuli and tissue damage near the primary afferent nerve terminals [6].

Primary DRG neurons culture has been a useful model system for investigating sensory physiology and putative nociceptive signaling [7]. ATP-induced intracellular calcium concentration ($[Ca^{2+}]_i$) transients in cultured DRG neurons have been used to model the response of nociceptors to painful stimuli. Visceral nociception and nociceptor sensitization appear to be regulated by P2X3 and TRPV1. Sensory neurons response to ATP and capsaicin suggest that visceral afferent nociceptors can be modulated by Ca^{2+} changes at a new site at the level of primary afferent neurons.

Several lines of evidence indicated that there is a close relationship between nerve fiber density and associated pain. There is a significant increase in nerve fiber density in women with inflammation who reported pelvic pain, suggesting these nerve fibers may play an important role in the mechanisms of pain generation. Accumulating literatures described that Substance P (SP) presents is involved in the inflammatory and pain responses, suggesting a possible role of SP nerve fibers in the generation of pain related sensations. SP, which is synthesized and contained in 20–30% of DRG neurons, is involved in the transmission of nociceptive information to the central nerve system. SP is contained primarily in, and co-released from, small-diameter primary afferent fibers on noxious stimulation. Activation of nociceptive C and Aδ primary afferent fibers by electrical, chemical, or mechanical stimulation has been reported to release SP. Visceral nociceptive C-fibers can be activated by SP, representing an endogenous system regulating inflammatory, immune responses, and visceral hypersensitivity. SP afferent fibers play an important role in the pathogenesis of visceral hyperalgesia, suggesting critical role of SP in regulation of visceral nociception. ATP is a peripheral mediator of pain, which contributes to the activation of sensory afferents by activating ATP receptors following inflammation or nerve injury. It may correlate with SP release and play an important role in modulating nociception in primary sensory neurons.

The response properties of pelvic extrinsic primary afferent nerves play a significant role in etiology of many functional disorders Hypersensitivity of visceral mechanoreceptors could result from excessive production of modulatory neurotransmitters. In addition to direct stimulation of stretch activated channels on primary afferent neurons located in dorsal root ganglia (DRG), chemicals produced by different target cells that respond to inflammation also modulate the nociception. The incidence of persistent, episodic, and chronic visceral pain is more prevalent in females, which suggests hormonal regulation of visceral nociception. Despite extensive research on the properties of pelvic and splanchnic afferent nerves, little is known about the mechanisms underlying normal and pathological signal transduction pathways underlying many functional diseases. Considerable efforts were made by the scientific community and the pharmaceutical industry to develop novel pharmacological treatments aimed at chronic visceral pain, but the traditional approaches to identify and evaluate novel drugs have largely failed to translate into effective therapeutic treatments of functional diseases. DRG neurons in short-term culture retain the expression of receptors (P2X and TRPV1) which mediate the response to putative nociceptive signals. In general, depending on the channel activation activity of primary afferent neurons may result in hyperpolarization, depolarization or primarily Ca^{2+} influx. DRGs transmit information about chemical or mechanical stimulation from the periphery to the brain. They continue to respond to different agonists mimicking in vivo activation. Nociceptors are small to medium size DRG neurons whose peripheral processes detect potentially damaging physical and chemical stimuli. The peripheral sensitization of nerve fibers is transient depending on the duration of stimuli and presence of visceral inflammation. An important advantage is that these neurons can be studied apart from endogenous signals. Available data clearly showed the new role of nociceptors in pathophysiological aspects of chronic pelvic pain through changes in Ca^{2+} signaling and potential way of designing future therapies.

A multicomponent conceptual model of functional disorders involves physiologic, cognitive and behavior factors will be necessary for developing new therapies. The different systems such as neuroendocrine regulation and autonomic response will affect nociceptive systems. From a public health perspective, a substantial impact on our knowledge of nociceptive functional diseases will help achieve a deeper understanding of data presented in clinical aspects of these symptoms. Only a thorough understanding of the mechanism implicated in these phenomena can truly contribute to the designing of new and more efficient therapies.

REFERENCES

[1] Malykhina, A. Neural mechanisms of pelvic organ cross-sensitization, *Neuroscience,* 2007, 149, 3: 660-672.
[2] Giamberardino, M and Costantini A. et al. Viscero-visceral hyperalgesia: characterization in different clinical models. *Pain,* 2010, 151, 2: 307- 322.
[3] Pezzone, M., Liang R., et al. A model of neural cross-talk and irritation in the pelvis: implications for the overlap of chronic pelvic pain disorders. *Gastroenterology,* 2005, 128, 7: 1953- 1964.
[4] Chaban V. Visceral pain modulation in female primary afferent sensory neurons, *Current Trends in Neurology,* 2015, 9: 111–114.

[5] Petruska, J., Napaporn J., et al. Subclassified acutely dissociated cells of rat DRG: histochemistry and patterns of capsaicin-, proton-, and ATP-activated currents. *Journal of Neurophysiology,* 200, 84, 5: 2365- 2379.

[6] Burnstock, G. Purine-mediated signalling in pain and visceral perception. *Trends in Pharmacological Sciences,* 2001, 22, 4: 182-188.

[7] Chaban V. *Unraveling the Enigma of Visceral Pain,* 2016, Nova Publishers, N.Y. 90p.

Chapter 10

CALCIUM SIGNALING AND REGULATION OF BRAIN FUNCTION

ABSTRACT

The solid evidence in published literature supports the role of calcium signaling not only in neuronal and glial cells but also in specific signaling cascade participating in the general brain functions from synaptic transmission to neuroplasticity and memory changes.. This signaling can be separated into the immediate such as Ca^{2+} binding to appropriate proteins and channels and sustained effects modulating gene expression and processing of the synthesized molecules. Insufficient Ca^{2+} causes neurons to produce action potentials spontaneously sending messages around the brain, which leads to "confusion" states. Common neurological diseases such as Alzheimer's disease, age- related dementia, amyotrophic lateral sclerosis, diabetic neuropathy and others all have been related to Ca^{2+} homeostasis. Modulation of gene expression, cell differentiation, apoptosis and autophagy exert extensive long- lasting effects that controlled by $[Ca^{2+}]_i$ perturbations. Advancements in understanding the Ca^{2+} signaling circuitry and the availability of specific molecular and genetic tools to probe Ca^{2+} circuits will allow us to determine the brain signaling, thereby revealing key mechanisms and new therapeutic strategies.

ROLE OF PRIMARY AFFERENT NERVES IN THE MECHANISMS OF CALCIUM NEUROTRANSDUCTION ASSOCIATED WITH CHRONIC PAIN

Pelvic nerve afferent fibers innervating the visceral organs of the lower colon have been well characterized [1]. In general, during colonic distension, a large number of pelvic EPANs show static levels of discharge. Stretches that lead to the opening of stretch-activated (SA) channels on the plasma membrane lead to the selective or non-selective opening of different cation and anion channels in nodose ganglia and DRG neurons. Thus, depending on the cell type and channel type, EPANs activation may result in hyperpolarization, depolarization, or primarily calcium influx. The function of SA channels in the plasma membrane differs between various cell types. Influx of Ca^{2+} may repolarize the plasma membrane via activation of voltage gated calcium channels (VGCCs) and thus influence adaptation rates of sensory neurons during their stimulations. Chronic pelvic pain is defined as pelvic pain in the same location for at least 6 months, affects up to 25% of women at some time in their lives and can result in dysmenorrhea, dyspareunia, menstrual irregularities, back pain, gastrointestinal and genitourinary symptoms and reduced fecundity. Some of the most common causes of chronic pain in gynecology are endometriosis, vulvodynia and abdominal wall and pelvic floor myalgia and neuropathy. Most women with complaints of pelvic pain will undergo laparoscopy to both diagnose and treat these diseases, but laparoscopy is often is unsuccessful due to lack of intraperitoneal pathology or altered pain processing. Pain out of proportion to identifiable pathology is the most immediate and dramatic consequence of disorders associated with chronic pelvic pain and is responsible for a highly negative impact on quality of life and substantial workforce loss. Due to the alarming situation and unmet need, many countries have launched a call for more focused research on improving the treatment of chronic pain syndrome. Pain is *subjective feeling* and thus difficult to be standardized and parameterized for scientific analysis. Often the etiology of chronic pain is not clear, as there are many symptoms of the reproductive system, gastrointestinal and urinary tracts, musculoskeletal, neurological and psychological systems that often co-occur in the same patient. Pain is also the *subjective experience* which can occur, particularly in its chronic manifestation, in the absence of physiological findings that explain its persistence or severity. Because of the inherent subjectivity of pain, there is a wide disparity among individuals in the way that they experience the pain generated by what seem to be similar stimuli. Pain is a *complex and individual experience* that is often difficult for patients to fully describe using conventional pain intensity.

Most of the current literature pertains to specific chronic pain syndromes defined by medical subspecialties. These include: Irritable Bowel Syndrome (IBS: gastroenterology); Chronic Pelvic Pain Syndrome (gynecology); Painful Bladder Syndrome (PBS: urology); fibromyalgia (rheumatology) and others. Many reports described the substantial overlaps between two or more of these syndromes [2]. Moreover, clinical presentations of functional syndromes lack a specific pathology in the affected organ but may respond to a viscero- visceral cross-sensitization in which increased nociceptive input from an inflame organ (i.e., uterus) sensitizes neuron that receive convergent input from an unaffected organ- (i.e., colon or bladder). The site of visceral cross-sensitivity is currently unknown.

Viscero- somatic and viscero- visceral hyperalgesia and allodynia lead to the perception of pain spreading from an initial site to adjacent areas. Patients with chronic pain may at first have only one source of pain in the pelvis, but numerous mechanisms involving the central and peripheral nervous systems may result in the development of painful sensations in adjacent organs, such as being associated with lower colonic pain Similar to other chronic diseases, a multicomponent conceptual model of pelvic pain, which involves physiologic, cognitive and behavior factors will be necessary for developing new therapies. The different systems such as neuroendocrine regulation, pain modulation, and autonomic response will affect ascending somatosensory and descending efferent systems (Figure 10.1).

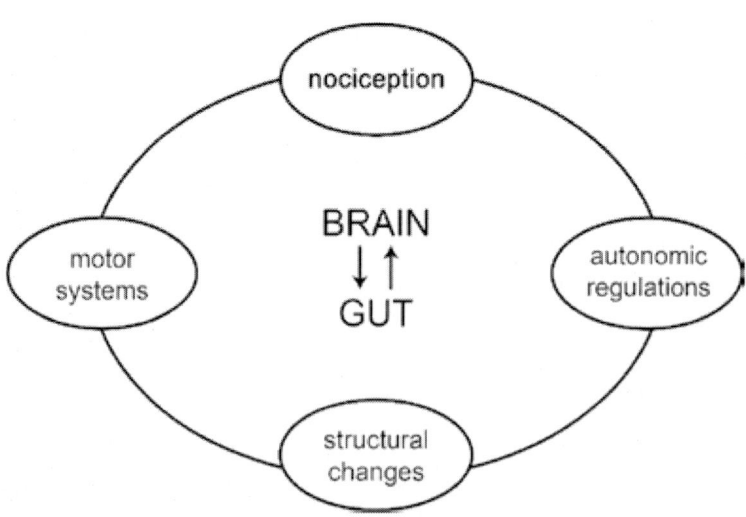

Figure 10.1. Different regulatory mechanisms involved in chronic pain [3].

Pain is *subjective feeling* and thus difficult to be standardized and parameterized for scientific analysis. Often the etiology of chronic pain is not clear, as there are many symptoms of the reproductive system, gastrointestinal and urinary tracts, musculoskeletal, neurological and psychological systems that often co-occur in the same patient. Pain is also the *subjective experience,* which can occur, particularly in its chronic manifestation, in the absence of physiological findings that explain its persistence or severity. Because of the inherent subjectivity of pain, there is a wide disparity among individuals in the way that they experience the pain generated by what seem to be similar stimuli. Pain is a *complex and individual experience* that is often difficult for patients to fully describe using conventional pain intensity.

Chronic pain patients have significantly more depression, psychological and somatic complains and more often give a history of physical, sexual or emotional abuse or trauma. Chronic pain adversely affects mood, social and professional life and general well- being. These quality of life issues can affect severity of pain, degree of impairment resulting from a painful condition and success of pain treatment modalities in alleviating pain. The treating clinician is often tempted to take a unidimensional approach and focus on one organ system and ignore the psychological and behavioral manifestations of the chronic pain. Chronic pain syndrome implies the pain symptoms are not just a manifestation of a peripheral pelvic abnormality, but a complex entity in which traditional therapy is less effective due to altered central nervous system processing of pain signals. Sensitization of cell bodies of primary afferent neurons located within the DRG ganglia innervating the different organs may be through the release of nociceptive transmitters such as ATP and/or substance P within the ganglion. Both act through changes of $[Ca^{2+}]_i$ signaling within the DRGs.

CALCIUM PATHWAYS IN COMMON BRAIN PATHOLOGIES

Calcium changes are almost always involved in pathology of brain function. Most common pathologies include those associated with aging such as Alzheimer's disease, dementia, Parkinson's disease, amyotrophic lateral sclerosis, diabetes, and different neuropathies between the other brain-associated diseases. The age-dependent changes in $[Ca^{2+}]_i$ signaling depends on alteration on the resting $[Ca^{2+}]_i$ and transmitter- induced changes in cytoplasmic Ca^{2+} concentration. Indeed, the age-related disturbance in resting $[Ca^{2+}]_i$ was found in DRG primary sensory neurons isolated from old (30 month) rats as well as from granule neurons in cerebellum in aged mice The reduction of $[Ca^{2+}]_i$ amplitude is associated with Na^+/Ca^{2+} exchange and Ca^{2+} pumps, mitochondrial calcium uptake as well as cytoplasmic Ca^{2+} buffering [4]. Ischemia also leads to excessive Ca^{2+} neuronal influx. Certainty, these changes represent a complex phenomenon that involves many intracellular systems regulating $[Ca^{2+}]_i$ homeostasis that may explain increased vulnerability of aging central nervous system to neurodegeneration.

In Alzheimer disease (AD) etiology development of β- amyloid in neurons which associated with neurodegeneration, lead to rapid increase of $[Ca^{2+}]_i$ leading to apoptosis. However the presence of Ca^{2+} channels blockers significantly attenuated the progression of molecular changes of with this disease. One of he most sensitive part of the brain to neurodegeneration such as hippocampus and closely associated neocortex associated with memory express high density of Ca^{2+}- sensitive glutamate receptors [4]. New evidence for a central role of Ca^{2+} in neurodegeneration includes the temporal and spatial relationships of molecular mechanisms that regulate neuronal Ca^{2+} ions, changes of this ion in different subcellular compartments and alterations in Ca^{2+} signaling that affect the performance of neurons in a healthy state or during aging and in disease [5].

Parkinson's disease (PD) is the second most common neurodegenerative disease after AD.

During development of PD dopaminergic (DA) neurons in Substantia Nigra (SN) region generate spikes that maximizes Ca^{2+} entry and promotes slow autonomous rhythmic activity accompanied by slow oscillations in $[Ca^{2+}]_i$ that are caused by the opening of plasma membrane Cav1 Ca^{2+} channels and release of Ca^{2+} from endoplasmic reticulum [6]. This mechanism leads to the possibility that Ca^{2+} entry through Cav1 channels and Ca^{2+}-dependent changes of mitochondrial metabolism damage SNc DA neurons linked to PD (i.e., aging, mutations or metabolic malfunction). High cytosolic Ca^{2+} and mitochondrial stress could drive neurodegeneration.

Glutamate toxicity and dysfunction of ligand-gated Ca^{2+} channels can also cause neuronal damage associated with amyotrophic lateral sclerosis (ALS) where spinal motoneurons are damaged. Increased glutamate level may lead to an excessive Ca^{2+} influx in ALS and increased Ca^{2+} level may lead to misfolding of proteins facilitating their toxic aggregation [7]. Furthermore, an increased cytosolic Ca^{2+} level may impair the buffer systems of the endoplasmic reticulum causing Ca^{2+} leak. Therefore, Ca^{2+} dyshomeostasis can be linked to intracellular dysfunction, being one of the main processes associated with ALS. Different Ca^{2+}-release channels are localized to the cytoplasm such as ATP-gated P2X4; the member of the TRP (transient receptor potential) ion channels family, P/Q-type voltage-gated Ca^{2+} channels and the TRPs, TRPA1non- selective cation channels [8].

There is increasing evidence suggesting a cross- talk between Ca^{2+} homeostasis and oxidative stress (OS) cascade. OS modifies Ca^{2+} signaling proteins and reshape local and global Ca^{2+} amplitudes and kinetics associated with various intracellular signal transduction pathways. Stimuli induce the entry of external Ca^{2+} *via* TRP channels, and G-proteins coupled receptors (GPCR) triggering release of internal Ca^{2+} from the ER by formation of second messengers that open channels of receptors for inositol trisphosphate: $InsP_3R$ and ryanodine: RyR. Over 90% of GPCR are expressed in the brain. In the CNS, GPCR function primarily as mediators of slow neuromodulators rather than of fast neurotransmitters in normal brain function. RyR and $InsP_3R$ are localized in the cytoplasm and the internal nuclear membrane and predominantly operated by second messengers. The neurons that ncrease in extracellular glutamate concentration and Ca^{2+} homeostasis sustain long-term changes in Ca^{2+} signaling that are prominent features of the epilepsy [9]. Many Ca^{2+}-dependent processes, such as susceptibility synaptic plasticity, to neurotoxicity, long-term potentiation (LTP) and long-term depression (LTD) are altered during disease or with age [10].

Although the initial reports regarding a role of calcium ions in neuroprotection were published about 30 years ago, we know today that Ca^{2+} entry into the cell through glutamate-sensitive NMDA, AMPA as well as many other transporters such as TRP family, CaV1 channels and acid-sensitive channels (ASIC). Within the cell most important sources of Ca^{2+} effects are ER, mitochondria and plasmalemma. Further, activation of protein kinases such as PKA and PKC, proteases, calpains, calcimeurin, calmodulin, nitric oxide synthase (NOS) play major role in signal transduction leading to either neuroprotection or disease.

Many Ca^{2+} targeting drugs have been demonstrated to protect neurons in pre-clinical trails, however only few have been successful in humans. Therefore, there is increased need to obtain more rigorous scientific evidence for the suggested mechanism of action on Ca^{2+} signaling in neuroprotection during critical clinical evaluation for potential new therapeutic interventions,

CALCIUM AS THERAPEUTIC TREATMENT IN THE NERVOUS SYSTEM

Most of available scientific and clinical data support role of calcium signaling as powerful modulator of almost all cellular functions through immediate or sustained actions. This positively charged ion mainly contained in bone and its small amount regulated by calcitonin, parathyroid hormone (PTH) and vitamin D in the circulation. Vitamin D is a group of steroid molecules that triggers intestinal absorption of Ca^{2+} and PTH mostly controls Ca^{2+} plasma level. About half of the calcium in the body is bound to albumin but only free or ionized calcium participates in numerous enzymatic reactions, release of neurotransmitters and neuronal excitability. Hypocalcaemia leads to alterations in neuromuscular excitability, hypotension and requires calcium supplementation. Hypercalcemia will decrease neuromuscular excitability leading to muscle weakness. Treatment targets excretion of calcium or inhibition of bone breakdown.

Calcium mostly stored in the bone tissue but as we age, we absorb less and less calcium from our diet, causing our bodies to take more and more calcium from our bones. In respect to nervous system calcium signaling plays a significant role in regulating multiple aspects of neuronal activity: functional Ca^{2+} spikes regulate neurotransmitter release, whereas Ca^{2+} waves regulate the rate of axon potential, indicating neuronal specificity. Overall, neuronal calcium signaling provides researchers and clinicians better understanding of the function of neuronal circuits in healthy states. At the same time, understanding of different mechanisms of calcium alterations provides significant insight into the pathogenesis of such disorders as Alzheimer's disease, Parkinson's disease, Huntington's disease, amyotrophic lateral sclerosis and others characterized with neuronal degeneration. L-type calcium channel antagonists have been used in bipolar disorder for over 30 years and about 45 percent of all person and 70 percent of older respondents reported calcium intake from supplements in two most common forms: calcium carbonate and calcium citrate [11]. Bioavailability increased when Ca^{2+} is solubilized and inhibited with binding of Ca^{2+} to other compounds by forming insoluble salts. Calcium citrate is less dependent than calcium carbonate on stomach acid for absorption. Interestingly, high Na^+ intakes increase Ca^{2+} excretion into the urine.

Therapeutically, calcium is used to prevent and reduce pain following chemotherapy. Calcium is given to patients with very low blood calcium levels and to patients with high potassium levels. In contrary, hypercalcemia can cause neurological symptoms, such as depression, amnesia, and nervousness. Severe cases can cause delirium and coma. Common diuretics can produce hypercalcemia by underexcretion of Ca^{2+}. Adults require a balanced Ca^{2+} intake from diet or medications for healthy body function. Hypercalcemia or hypocalcemia both cause serious adverse effects. The National Academies of Sciences Food and Nutrition Board recommends 1,000 to 1,200 mg Ca^{2+} per day for most adults. Older adults, and women who are pregnant or breastfeeding and some adolescents may require additional Ca^{2+} supplementation based on the patient's evaluation by health care provider [12].

Normal calcium signaling is important for functioning of all cells particularly neurons coupled to control mechanisms of the body. Ca^{2+} – induced intra- and intercellular signaling triggers biochemical cascade linking cellular physiology to metabolic changes and control of gene expression, cell differentiation and apoptosis within the cells and tissues. Therefore, calcium plays a unique and special role in maintaining functional state of living matter.

REFERENCES

[1] Mayer E., Labus J., Tillisch K., Cole S., Baldi P. Toward a systems view of IBS. *Nature Review Gastroenterology and Hepatology*, 2015, 12, 10: 592- 605.

[2] Malykhina A. Neural mechanisms of pelvic organ cross-sensitization. *Neuroscience*, 2007, 149, 3: 660– 672.

[3] Chaban V. Irritable Bowel Syndrome: Functional Gastrointestinal Disease Regulated by Nervous System In: *"Irritable Bowel Syndrome- Novel Concepts for Research and Treatment"* (Chaban V,- Ed.), InTech, 2017, 85p.

[4] Kostyuk P. and Verkhratsky A. *Calcium Signaling in the Nervous System,* Wiley, 1995, 206p.

[5] Alzheimer Association Calcium Working Group. Calcium Hypothesis of Alzheimer's Disease and Brain Aging: A Framework for Integrating New Evidence Into a Comprehensive Theory of Pathogenesis. *Alzheimer and Dementia,* 2017, 3, 2: 178-182.

[6] Bean B. The action potential in mammalian central neurons. *Nature Review Neuroscience,* 2007, 8: 451- 465.

[7] Leal S., Cardoso I., Valentine J., Gomes C. Calcium ions promote superoxide dismutase 1 (SOD1) aggregation into non-fibrillar amyloid: A link to toxic effects of calcium overload in amyotrophic lateral sclerosis (ALS)? *Journal of Biological Chemistry,* 2013, 288: 25219– 25228.

[8] Tadeschi V., Petrozziello T and Secondo A. calcium dyshomeostasis and lysosomal Ca^{2+} dysfunction in amyotrophic lateral sclerosis. *Cells,* 2019, 8, 10: 1216- 1231.

[9] Nagarkatti N. Deshpande. DeLorenzo R. Development of the calcium plateau following status epilepticus: role of calcium in epileptogenesis. *Expert Review in Neurotherapeutics,* 2009, 9: 813- 824.

[10] Foster T. Calcium homeostasis and modulation of synaptic plasticity in the aged brain. *Aging Cell.* 2007, 6: 319- 325.

[11] Bailey R, Dodd K, Goldman J, Gahche J, Dwyer J, Moshfegh A, Sempos C, Picciano M. Estimation of total usual calcium and vitamin D intake in the United States. *Journal of Nutrition,* 2010, 140, 4: 817–22.

[12] Ross A., Taylor C., Yaktine N., and Del Valle H. (Editors) *Dietary Reference Intakes of Calcium and Vitamin D.* The National Academies Press, Washington DC, 1114p.

ABOUT THE AUTHOR

Victor V. Chaban, PhD, MS
Professor of Medicine
Charles R. Drew University and University of California Los Angeles, USA

Dr. Victor V. Chaban is Professor of Medicine with dual appointments at Charles R. Drew University of Medicine and Science (CDU) and University of California Los Angeles (UCLA). Dr. Chaban completed his post-doctoral training in Neuroscience at UCLA and graduate studies in Clinical Research at CDU. Dr. Chaban is a member of United States National Institute of Health and several international study sections. Dr. Chaban serves as Editor In Chief for the International Journal of Research in Nursing. He is an author of: "Unraveling the Enigma of Visceral Pain" by Nova Publishers, N.Y. 2016, Editor of: "Irritable Bowel Syndrome: New Concepts for Research and Treatments" by InTech, London, 2017 and "Neuroplasticity: Insights of Neural Reorganization", InTech, London, 2018.

INDEX

A

action potential, xii, 5, 9, 33, 79, 88
afferent nerve, 74, 75
afferent neuron, xi, 2, 33, 37, 41, 47, 50, 63, 67, 69, 74, 75, 83
agonist, 13, 15, 17, 21, 23, 26, 42
albumin, 86
allodynia, 64, 72, 81
AMPA, 85
antibody, 56
astrocytes, v, 53, 54, 55, 57, 58, 59, 60, 61, 62
axon, viii, 86

B

back pain, 65, 73, 80
BAPTA, 3, 14, 23
bladder, 69, 70, 72, 81
bowel, 2, 50, 51, 63, 64, 69, 71, 72, 81, 88
bradykinin, 42, 74
brain, vi, xii, 53, 54, 70, 76, 79, 83, 85, 88

C

Ca2+ channels, vii, viii, 4, 10, 60, 83, 84
calcitonin, 86
cAMP, 50
cancer, 19, 21, 26, 27
central nervous system (CNS), xi, 31, 32, 33, 69, 72, 83, 85
chronic pain, 32, 68, 69, 71, 72, 73, 80, 81, 82
circulation, 86
colon, 39, 41, 49, 64, 67, 70, 80, 81

D

depolarization, viii, 4, 5, 32, 33, 76, 80
depression, xii, 64, 82, 87
diabetes, 83
diabetic neuropathy, 79
dorsal horn, 32
dorsal root ganglia (DRG), xi, 1, 2, 3, 4, 5, 7, 8, 9, 10, 11, 13, 14, 15, 16, 17, 19, 21, 22, 23, 24, 25, 26, 27, 28, 31, 32, 33, 34, 37, 38, 39, 40, 41, 42, 43, 44, 45, 46, 47,

48, 49, 61, 62, 63, 66, 67, 69, 74, 75, 77, 80, 83
dysmenorrhea, 65, 72, 73, 80
dyspareunia, 65, 80

E

endometriosis, 2, 67, 71, 72, 73, 80
estradiol, 7, 9, 10, 11, 13, 14, 15, 17, 19, 21, 22, 23, 24, 25, 26, 28, 38, 42, 44, 46, 48, 51, 53, 54, 57, 58, 59, 60, 62, 66, 67
estrogen, v, xii, 6, 7, 8, 9, 10, 13, 14, 15, 16, 18, 19, 33, 34, 36, 37, 45, 49, 51, 53, 54, 61, 62, 63, 66, 69, 70
excitability, 33, 61, 68, 73, 86

F

fibers, 2, 4, 5, 32, 49, 74, 80
fibromyalgia, 63, 69, 72, 81

G

ganglia, 6, 18, 33, 39, 63, 67, 80, 83
gender differences, 34, 69
glial cells, viii, xii, 54, 79
glutamate, v, xii, 7, 8, 9, 10, 11, 34, 61, 67, 84, 85

H

hippocampus, 84
hydrogen, 27
hyperalgesia, 10, 33, 41, 49, 50, 64, 72, 73, 75, 76, 81
hypersensitivity, 5, 10, 32, 69, 75

I

iatrogenic, 65

inflammation, 1, 5, 6, 33, 34, 41, 47, 48, 66, 67, 69, 74, 75
Inflammatory Bowel Disease (IBD), 71
Inflammatory Bowel Syndrome (IBS), 41, 51, 63, 64, 67, 71, 81, 87
interneurons, 32
IP3, 60

L

long-term depression (LTD), 85
long-term potentiation (LTP), 85
Long-term potentiation (LTP), 85
luteinizing hormone (LH), 54, 61

M

microglia, 54
mitochondria, 85
myalgia, 71, 80

N

nerve fibers, 74, 76
neuritis, 47
neuropeptides, 27, 70
neurotransmitters, 10, 75, 85, 86
NMDA, xii, 2, 6, 67, 85
nociception, v, vi, xii, 1, 2, 5, 6, 9, 10, 16, 27, 31, 33, 34, 35, 37, 49, 63, 69, 74, 75
nociceptors, 2, 5, 14, 16, 32, 33, 36, 50, 69, 70, 73, 74, 76

O

oligodendrocytes, 54
opioids, 5, 9

P

Painful Bladder Syndrome (PBS), 39, 56, 63, 64, 71, 81
protein kinase A (PKA), 50, 85
protein kinase C (PKC), 85
purinoceptors, 16

R

ryanodine, 85

S

sensitization, v, 1, 2, 4, 5, 27, 31, 32, 33, 34, 49, 67, 68, 69, 70, 73, 74, 76, 81, 83, 87
sensory neurons, xi, 1, 9, 10, 13, 18, 21, 24, 35, 37, 45, 46, 49, 51, 68, 72, 74, 75, 76, 80, 83

signal transduction, viii, xi, 8, 27, 34, 64, 75, 85
synaptic plasticity, 54, 85, 88
synaptic transmission, 79

U

urinary bladder, 64, 66
uterus, 39, 41, 49, 67, 72, 81

V

visceral, v, vi, xiii, 1, 2, 5, 6, 10, 16, 27, 31, 32, 33, 34, 36, 37, 38, 39, 41, 42, 43, 44, 45, 46, 47, 48, 49, 51, 63, 64, 65, 66, 67, 68, 69, 70, 71, 72, 73, 74, 75, 76, 77, 80, 81
visceral pain, 2, 32, 33, 37, 41, 49, 50, 51, 63, 64, 65, 66, 67, 68, 69, 72, 73, 75, 76